THE WIRELESS SPECTRUM:
THE POLITICS, PRACTICES, AND POETICS
OF MOBILE MEDIA

The Wireless Spectrum

The Politics, Practices, and Poetics of Mobile Media

EDITED BY BARBARA CROW, MICHAEL LONGFORD, AND KIM SAWCHUK

UNIVERSITY OF TORONTO PRESS
Toronto   Buffalo   London

© University of Toronto Press Incorporated 2010
Toronto Buffalo London
www.utppublishing.com
Printed in Canada

ISBN 978-0-8020-9893-1

Library and Archives Canada Cataloguing in Publication

The wireless spectrum : the politics, practices, and poetics of mobile media /
edited by Barbara Crow, Michael Longford, and Kim Sawchuk.

(Digital futures)
Includes bibliographical references.
ISBN 978-0-8020-9893-1

1. Wireless communication systems – Social aspects. 2. Interpersonal communication – Technological innovations – Social aspects. 3. Communication
and culture. I. Crow, Barbara A., 1960–   II. Longford, Michael III. Sawchuk,
Kim IV. Series: Digital futures

HM851.W557 2010   303.48'33   C2010-900381-0

University of Toronto Press acknowledges the financial assistance to
its publishing program of the Canada Council for the Arts and Ontario
Arts Council.

 Canada Council   Conseil des Arts   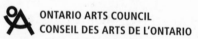 ONTARIO ARTS COUNCIL
for the Arts   du Canada   CONSEIL DES ARTS DE L'ONTARIO

This book has been published with the help of a grant from the Canadian
Federation for the Humanities and Social Sciences, through the Aid to
Scholarly Publications Programme, using funds provided by the Social
Sciences and Humanities Research Council of Canada.

University of Toronto Press acknowledges the financial support for its
publishing activities of the Government of Canada through the Book
Publishing Industry Development Program (BPIDP).

Contents

Acknowledgments

We would like to thank the following organizations, which generously funded the development of these papers: Canadian Heritage New Media Funds, Concordia University, York University, Ontario College of Art and Design, Banff New Media Institute, Société des arts technologiques / Society for Arts and Technology, and Centre interuniversitaire des arts médiatiques (CIAM).

This collection would not have been completed without the excellent work of graduate students Linnet Fawcett, Andrea Zeffiro, Samantha Moonsammy, Candice D'Souza, Alison Harvey, Neil Barratt, Ganaele Langlois, Sanja Obradovic, and, in particular, Janice Leung. Janice worked tirelessly formatting and administering the details of the book's production – thank you so much!

Finally, to our editors, Siobhan McMenemy and Ryan Vanijstee, copy editor Ken Lewis, and managing editor Frances Mundy, and our authors, thank you for your patience and important contribution to imagining the burgeoning wireless technologies.

THE WIRELESS SPECTRUM:
THE POLITICS, PRACTICES, AND POETICS
OF MOBILE MEDIA

1 Introduction

KIM SAWCHUK, BARBARA CROW,
AND MICHAEL LONGFORD

The Wireless Spectrum: The Politics, Practices, and Poetics of Mobile Media
explores how the proliferation of wireless communication technologies
modifies individual and public life, transforms our experience of space,
time, and place, and reshapes our day-to-day interactions. The authors
included in this anthology examine the multifarious uses of a range of
mobile media – cell phones, personal digital assistants (PDAs), museum
audio guides, and Wi-Fi or Bluetooth-enabled laptop computers – being
deployed by artists, activists, citizens, and consumers in innovative and
unexpected ways. These reflections are timely, as we are at a key moment in the proliferation and promotion of these devices.

In the last decade, the use of cell phones has increased fourfold in
Canada. In 1985 there were 6,000 subscribers to cellular service, in
1995 there were 2,584,387, and in 2007 this number had skyrocketed
to 19.3 million (CWTA 2007). The industry predicts an increase of
20 per cent in the cell-phone market in the next year alone. Spectrum
for 3G and 4G technologies are being auctioned for millions of dollars,
and media convergences are capitalizing on the technical capabilities of
wireless communications for the next generation of Internet, cellular
phones, and PDAs. Within this volatile context of economic investment,
contextualization, critique, and creative interventions are necessary to
ensure that segments of the electromagnetic spectrum are kept public,
accessible, and free of burdensome fees that hinder the non-commercial
potential of these technologies.

The Wireless Spectrum: The Politics, Practices, and Poetics of Mobile Media
is subdivided into four parts – 'Spectral Genealogies,' 'Mobile Practices,'
'Locative Media,' and 'Wireless Connections'– pinpointing some of the
major issues in the interlocking intellectual terrain of current studies of

mobile, wireless communications. The contributors include artists, academics, and renegade developers. As a whole, the articles interrogate how new mobile forms of community and practices disturb the corporate and industry agendas that so often guide the purchase and distribution of these technologies.

Part I, 'Spectral Genealogies,' focuses on some of the historical and conceptual antecedents of both wireless communications and the idea of mobility. The contributions of Will Straw, Kim Sawchuk, and Jennifer Gabrys remind us, theoretically and philosophically, how the 'new' continues to rely on the 'old' media. By locating these new communications technologies in relation to existing media histories (Sconce 2000, Manovich 2001, Chun and Keenan 2006, Gitelman 2006), the articles in this section implicitly connect present research concerns to the past, allowing for a nuanced discussion of present developments technically and socially.

Part II, 'Mobile Practices,' puts the spotlight on the users of these technologies. The articles in this section, by Anne Galloway, Judith Nicholson, Sandra Buckley, and Darin Barney, explore how the practices of citizens, artists, and consumers who are using mobile technologies both transform and signal the changing dynamics of private to public places. This focus on examining user-integration and innovation, taking into account specific social and cultural contexts, is critical as mobile technologies increasingly permeate all facets of everyday life (Ling 2004, Goggin 2006, Katz 2006, Castells et al. 2007, Caron and Caronia 2007).

The contributions and interventions of artists to the field of telecommunications are made explicit in Part III, 'Locative Media,' whose contributors include Minna Tarkka, Kajin Goh, Michael Longford, and Barbara Crow. This term, which was coined in the late '90s, has been adopted by artists, activists, and community-based technologists interested in alternative cartographies and the potential of public authoring to augment and expand connections between local communities as increasingly mobile devices are able to access the Internet. In the tradition of locative media, artists have utilized a range of devices, from GPS systems and Wi-Fi networks to Bluetooth-enabled phones, to create new artworks, community maps, or mobile gaming experiences that engage and make visible the particularities of local spaces and place (Hoete 2003, Lane 2004, Tuters and Smite 2004). Many locative media artists creatively deploy commercial technologies for non-commercial uses in an effort to render visible the ubiquity of wireless communication systems. By appropriating and redeploying these technologies away from their military origins, they allow individuals, groups, and communities to reinsert themselves virtually into public spaces that are increasingly being co-opted by commercial interests.

Finally, Part IV, 'Wireless Connections,' grapples with issues of access, local engagement, and community-based Wi-Fi ventures. This group of articles is primarily grounded in community and activist media practice. It includes a reprint of the Wireless Commons Manifesto, and contributions by Andrew Herman, and Alison Powell. Like earlier wireless communication advocates for radio and ham radio, these ad hoc individuals and groups of community technologists have been keen to politicize and make visible the next generation of unlicensed spectrum, to make it 'free,' by developing and connecting user-friendly and accessible wireless communications technologies and open-source software platforms to various community-based Internet projects.

The Wireless Spectrum: The Politics, Practices, and Poetics of Mobile Media offers a unique contribution to a growing academic discourse emerging in communications, sociology, urban geography, and art criticism on the cultural practices shaping the use of mobile wireless technologies, expanding the purview of ICT (information-communication-technologies) research, and questioning some of its fundamental assumptions. Within ICT studies, there has been a tendency to focus on access to the Internet or the World Wide Web in locational ways: one works from home or the office or school, which are imagined as 'wired' together. While communication technologies may break down these discrete borders, the radical rupture of space and location introduced by wireless hand-held devices needs further examination for the specific ways that it creates invisible grids or 'clouds of data,' as Jennifer Gabrys so eloquently describes them, which hover over cities and re-map our understanding of place. In this way, we advocate for a distinction within ICT studies between ICT and WCT, or wireless communication technologies.

Wireless and mobile communications is also an emergent sub-field of new media research. Many Canadian artists and institutions have paired up with academic researchers, and sometimes industry, to forge new productive connections. Notable here are the contributions of Canadian artists such as Murmur, Marc Tuters, and Jason Lewis, who explicitly make new media works for transitory public spaces, such as street-level electronic billboards, and whose artistic practices question the nature of private-to-private messaging by facilitating private-to-public messaging. Projects initiated out of the Banff New Media Institute, including the Mobile Digital Commons Network, have intersected with international movements such as the work of the m-cult lab in Finland and England's Mobile Bristol project. This collection puts these kinds of artists' projects into conversation with academic and activist work, and foregrounds the international impact and circulation of Canadian scholarship in this emergent and highly interdisciplinary sub-field of communication and cultural studies.

The conceptual focus of the collection is the term 'spectrum,' which acts as a touchstone and overarching theme binding the four sections. None of the aforementioned devices could function without the scientific knowledge and complex infrastructure that has been developed to harness and profit from the electromagnetic energy that naturally occurs across the universe. As Gordon Gow and Richard Smith (2006) rightly contend, 'It is electromagnetic energy that enables us to broadcast radio signals over the air. It exists everywhere in the universe, it is invisible and it works in ways that are still somewhat of a mystery to scientists' (6). 'Radio spectrum' is the precise term that scientists, engineers, and policy-makers use to classify this undifferentiated miasma of electromagnetic energy into discrete bands of frequencies, which can then be allocated for different purposes, such as radio, telephone, television, and Wi-Fi. This present meaning of spectrum is imbued with a rich and suggestive past.

Etymologically, 'spectrum' is derived from the Latin word for image or apparition. This ghostly residue of what has become a scientific idea has had a fascinating if somewhat marginal place within communications and cultural studies. Avital Ronell's *The Telephone Book* (1989), John Durham Peters's *Speaking into the Air* (1999), and Jeffrey Sconce's *Haunted Media* (2000) articulate the history of communications technologies with respect to the desire to communicate with the dead. Jennifer Gabrys's article for the present volume, 'Atmospheres of Communications,' poetically evokes this history in its descriptions of the ethereal clouds of data that we meander through daily. As Ronell's, Peters's, and Sconce's work indicates, these spiritualist and otherworldly qualities of electromagnetism were never fully separate from the science and engineering of electronic signal processing and transmission. In modern electrical engineering, 'radio spectrum' refers to the complete range of frequencies or wavelengths that comprise electromagnetic energy. The term 'spectrum' now applies to any signal that can be decomposed into frequency components or bands. This energy has both electrical and magnetic properties that make different portions of the spectrum suitable for different purposes.

While spectrum, like the air we breathe, seems to exist in unlimited quantities, it is important to understand spectrum or these radio-wave frequencies as a renewable yet finite resource. It is renewable because as far as we know it does not erode or disappear with use; it is finite because it is unclear what the carrying capacity of a frequency may be, and for this reason it has been licensed nationally and internationally. Within the contemporary policy realm that governs the distribution of spectrum, two major classifications exist: licensed spectrum and unlicensed spectrum.

Licensed spectrum is regulated by governmental bodies that endeavour to ensure that radio frequencies are apportioned to minimize interference along a given channel; to create standards and conditions of compatibility between devices across nations and companies; and to decide who can broadcast on a given frequency. Within communications, the ability of companies to create devices that use the unlicensed spectral capacities of the frequency 802.11 have sparked debate over carrier and content. These are increasingly fractious as the convergence of mobile media platforms now allows for the distribution of user-generated content via the cell phone to sites on the World Wide Web such as YouTube, Myspace, or Facebook.

As we write this introduction, Rogers Communications has just begun a new promotion package for users to access YouTube through their mobile services, while Bell Canada Enterprises (BCE) has harnessed the popularity of Les Têtes à Claques to sell mobility to Quebeckers (Robertson 2008). By far the largest part of the spectrum is licensed by telecommunications companies such as BCE, Telus, and Rogers, in the case of Canada and elsewhere, who thus control access to the airwaves through extremely expensive agreements that make it virtually impossible for any but the largest companies to manoeuvre in this new invisible frontier of investment.

What is left for everyone and everything else is a small strip of communications: 802.3 for Ethernet use; 802.11 for Wi-Fi; 802.15 for Bluetooth; and 802.16 for Zigbee. These unregulated slices of spectrum were established in 1980 by the Institute of Electrical and Electronics Engineers (IEEE) for public use. There is no cost to the user to operate compatible devices in these sets of frequencies. It is this narrow window of spectrum that 'policy artists' such as Julian Priest, interviewed for this volume by Barbara Crow, have rendered visible to promote a better understanding of the politics of spectrum allocation to the public. Yet even the idea that these mobile devices are absolutely wireless is a fiction, for as many authors have pointed out, many of the features of any wireless device require access to land-line infrastructure, including hydroelectricity (Wilson 2008). Most cell-phone Internet access is reliant on public and private investment in fibre optics and underground and often invisible cables. From our purview, it is important to underscore the relationship between these telecommunications technologies and their residual reliance on other infrastructures. These technologies and systems that make it possible to live without wires or cables rely on the availability and licensing of spectrum.

It is this small portion of unlicensed, unregulated, freely available spectrum that has been mobilized by the vast majority of artists and activists who are part of the burgeoning community of Wi-Fi users engaged in the politics and economics of opening up more spectrum for public, non-commercial, not-for-profit uses, a politics discussed and critiqued by Andrew Herman in his contribution to this collection. As wireless communications media are becoming more abundant, Community Wireless Networks (CWNs) are committed to open-source software and infrastructure development in these frequencies and have been crucial in creating publicly available 'free' hotspots and mesh networks to provide free access to the Internet through Wi-Fi enabled computers or cellular phones. Alison Powell's article, based on original ethnographic research on Montreal's wireless community group Île Sans Fil, examines this ideal. This group of renegade engineers, artists, and activist volunteers have been an important force nationally and internationally as innovators and vocal advocates for the provision of public access to the Internet through wireless portals located in cafés and bars in the city. Despite this important mandate, Andrew Herman questions whether community wireless groups in Canada are pushing hard enough to open up more bandwidth, or whether they are making do with an unnecessarily small portion.

The initiatives of CWNs have had most of their success in Western nations such as Canada, the United States, and the United Kingdom. As more people around the world, particularly those in developing countries, desire access to the Internet through telephony, the potentials of CWNs will become increasingly important. It should be noted that these technologies are also limited by interference with other devices, such as microwaves, by physical requirements such as maintaining a line of sight between devices, as frequencies may be blocked by buildings and trees, and by disruption by weather.

While computing and engineering feats have made it possible to parse ever smaller parts of the radio spectrum, bandwidth is crucial not just for laptop computer users seeking access to the Internet from remote locations, but for all mobile devices, as more and more material is being uploaded and downloaded via these invisible channels of communication. This transformation of the cell phone from a two-way talking device to a machine for the uploading and downloading of user-generated content means that we are carrying powerful personal computers with us at all times. The functionality of the new phones allows for the uploading and downloading of information to Internet sites in a new ubiquitous computing and pervasive media environment for perambulatory users. The new 3G and 4G phones, such as the Apple iPhone, are also cameras, mp3 players, organizers, calendars, and alarm clocks in one small machine, and their interfaces are changing so that they look less like phones and more like new portable listening devices, such as the iPod. Three and 4G phones bring together a variety of media capabilities on one small mobile platform, including the ability to download films or songs, do on-line banking, and pick up e-mail or text messages. As Judith Nicholson describes in her contribution, the cellular telephone has become 'The Third Screen,' a term which makes explicit the phone's lineage from the screen practices of film, television, and computers.

This plethora of small hand-held devices vying for this precious and expensive resource is creating a major upheaval in the telecommunications and cultural industries in Canada and abroad. While the cell phone is only one of any number of portable devices that uses the spectrum, its sheer proliferation and transformation is astounding. Cell phones are the fastest growing consumer technology of the past one hundred years (Townsend 2004), and in 2007 cellular subscriptions surpassed land-line subscriptions for the first time globally. While its origins are in the military, commercial mobile telephony was provided in the United States as early as 1947. These early models were clunky and heavy because of the size of battery needed to operate them, and prohibitively expensive for all but the very rich. Cellular phones made their first widespread appearance in 1973, and these early units provided half an hour of talk time, weighed two pounds, and cost approximately $4,000 US dollars (*Popular Science* 1973). Today the lightest handsets weigh from three to six ounces. The sheer number of cellular telephones in Canada, indeed in the world, signals a remarkable shift in person-to-person communications within the broader media environment. The ubiquitous presence of the cell phone is changing the landscape of interactions in public space. This is visible through the growing presence of microwave towers atop office buildings, and audible in the one-way conversations many of us witness, participate in, and sometimes want to ignore in such public places as buses, streets, and cars.

While mobile technologies have become synonymous with cellular or mobile phones, cell phones are only the most visibly and audibly pervasive component of the emergent assemblage of immersive mobile technologies – and here a word on a minor distinction between mobility and portability in some of the literature on wireless communications and culture. For application designer Johan Hjelm (2000) there is a distinction to be made between mobility and portability. 'If you can't pick up the device and walk out of the building while maintaining contact and continuing to use the applications you were using, it is not really mobile' (Hjelm 2000, 1). Gow and Smith (2006) agree with this adage and develop their own spectrum between the poles of the fixed and the truly mobile. A telephone connected to a wall socket is a fixed device located in a particular space, while a truly mobile device, such as a cell phone, offers complete flexibility of movement between different spaces with continual contact. For these authors, this is the definition of 'true mobility' as opposed to portability (63). Laptops as portable devices are at another place on this continuum, between the truly mobile and the fixed in location.

In our 'sampling of the wireless spectrum,' we contend that one must understand these mobile and portable technologies as part of a larger network of all interoperable devices that can relay packets of information in bundles from one point to another. An analysis of portable, mobile, or wireless devices at a terrestrial level may include hybrid hand-held entertainment and gaming platforms, such as the latest Nintendo platform, the Wii. Many of these devices, which are connected to one another in a tangled assemblage of both wired and wireless infrastructure and to a variety of base stations, relay information over greater or lesser ranges. They include PANs (personal area network systems), which allow Bluetooth-enabled devices in close proximity to connect to each other, WANs (wide area networking systems), which cover larger areas, and LANs (local area network systems). While the cell phone is not the only mobile device, it is the perhaps the key technology involved in this multi-faceted transformation of social practices, with its new opportunities for cultural and political interventions such as those documented by Harold Rheingold in his early book *Smart Mobs* (2002), on the phenomena of flash mobs and the use of the phone for political action.

In short, the cell phone is only one of a range of interoperable portable devices with a long history in the social imaginary, of which Kim Sawchuk gives an account in her article 'Radio Hats, Wireless Rats and Flying Families' tracing the discourse of these terms in popular culture and images in the post-war period. Interestingly, one sees that the current promise that these technologies will enable and enrich the invisible lines of communication between individuals and families in new and exciting ways, and that they will create new opportunities for the delivery of cultural content, is part of a longer historical trajectory.

What is shifting is the line between carrier and content, enshrined in the governance of media technologies in Canada, and this shift is creating new corporate and governmental alliances. In Canada, traditional providers of content such as Bravo Television and the National Film Board (NFB) are eager to find new modes for disseminating their productions. The NFB, in conjunction with Bravo, has sponsored competitions and promotions of mini-movies that you can download to your phone. These private and public sector stakeholders are keen to engage with mobile technologies and to find innovative content that can take full advantage of the constraints of the small screens of the devices, but as well utilize a wider range of functions, such as uploading images and tagging them by location and time, in the case of GPS-enabled phones and the expanding Web services now being offered by Yahoo. These new services are being touted by some Web 2.0 officiandos as the latest indication that democracy is flourishing. These innovations in applications, many instigated from within the open source software movement, allow users to generate their own content and to be their own media producers. Yet such shifts are a two-edged sword, raising as they do issues of privacy, surveillance and data mining of citizens and consumers (Elmer 2004, Bradley 2005, *MUTE* 2005). When combined with global positioning satellite systems (GPS), mobile wireless technologies are an advertiser's dream for precise point-of-purchase marketing strategies delivered directly to your cell phone.

It is precisely these tensions – among shifts in notions of privacy, the potential of greater surveillance and tracking, and the intensification of consumerism and commodification – that artists working within the tradition of locative media – a tradition of artistic practice that examines movement in tandem with space – are addressing and negotiating. This tradition and these tensions are the subject of Minna Tarkka's essay, written from the context of Finland, where Nokia has pioneered the collusion and cooperation between artists and corporations, and increased cell-phone penetration to 106 per cent. Yet the creative potential of these technologies is that they may puncture a space with a presence from the past, or be used to augment the existing environment, as designers such as Kajin Goh and Michael Longford discuss with respect to their projects in Montreal's Parc Émilie-Gamelin.

While users may communicate on the move, these technologies can always and anytime pinpoint an emergent location or trajectory of movement. The introduction of these multi-faceted instruments into our lives is de-territorializing and re-territorializing (Deleuze and Guattari 1983) the traditionally understood relationship between the public and private in the mobile, wireless digital world we increasingly inhabit, creating new modalities of mediation and perhaps even a sense of phenomenological dislocation. It is this very sensation, in a contemporary gallery context and with respect to a very early mobile device – the audio guide – that is explored in 'Terminal City? Art, Information, and the Augmenting of Vancouver,' Darin Barney's offering to the present volume.

These devices also may create a new and distinct sense of the self. Anecdotal evidence suggests that when people leave their homes, they now bring three things with them: their keys, their wallets, and their cell phones. The pocket, the purse, the belt-holder: the cell phone, the PDA, or the portable PlayStation is not located at a station in a private home. The size of the screen and the size of the device locates it on an individual person. Advertisements selling the new status objects, cell phones or PDAs such as BlackBerrys or Treos, are supported by packages on offer ascribing a single phone and a single number to a single person. This is now reinforced by recent Canadian legislation connecting a cell number to individuals that can be moved with them even if they change carriers. These individualizing features embedded in the technologies create a relationship to these devices that means they are not easily shared among users. Perhaps less personal than a pair of shoes, underwear, or toothbrush, cell-phone agreements that provide service contracts to individuals do not encourage sharing, even as they appear to facilitate instantaneous access to other individuals. Having a cell phone means never having to have a fixed place with friends, as the time and place to meet can always be endlessly negotiated. In her contribution to this volume, Anne Galloway picks up on these questions of ubiquitous computing, tracing the historical shifts in the sense of public-ness and politics that it instantiates.

Drivers of cars, Bluetooth-enabled headsets blinking in their ears, encased in their vehicles, travel through public space further sequestrated in a bubble of private conversation as they meander down a busy city street. These new conditions for communications and their social uses are examined in the context of Japan by Sandra Buckley, who describes the emergence of thumb-tribes in Japan, as well as the decision by Japanese phone companies to keep the costs of communications exceedingly low in order to allow consumers to generate innovations in use. This is a practice distinctly different from North America, for example, where the high costs of phone services in the form of exclusive, lengthy contracts with service providers, have prohibited this type of activity. While cell-phone use has increased in Canada enormously, Canada lags behind other nations in terms of the proliferation of phones.

This condition of increased mobility accompanied by the transformation of a sense of a public-ness is described by Raymond Williams in his discussion of 'mobile privatization.' Within this context of use anywhere, anytime, and anyplace, mobile devices have created new objects for theoretical speculations and inaugurated what Will Straw describes in his contribution to this collection as 'the circulatory turn' in communications.

The articles collected in this volume are engaged in this new terrain of interdisciplinary research on mobile technologies in Canada and across the world, with the majority of its contributors providing a snapshot of the current conjuncture of mobile technology emergence. A unique contribution to these debates, *The Wireless Spectrum: The Politics, Practices, and Poetics of Mobile Media* brings together writers from communications, cultural studies, design, and new media, and collectively asks how wireless communications technologies present a challenge to standard conceptualizations of time, place, space, and location while transfiguring subjectivities and creating new forms of sociality and provocative aesthetic practices.

PART ONE

Spectral Genealogies

2 The Circulatory Turn

WILL STRAW

The Stable and the Variable

> La transmission des savoirs se joue entre l'ouverture et la clôture, entre condensation et dissémination, entre nomadisme et ancrage dans le territoire. (Bourassa 2005, 21)

Familiar images from the world of mobile, wireless technologies show us two individuals communicating with each other, via digital device, across shifting relationships of distance and directionality. (Think of any episode of 24, for example.) In these scenarios, people pursue their own, uncoordinated pathways of movement, against the backdrop of unwavering wireless networks. The fundamental dialectic of mobile communications is expressed here, in the open unpredictability of human action and the (relative) constancy of the network or infrastructure which makes communications possible. Indeed, the roving mobile user may seem the perfect embodiment of all those models of social action which set agency against structure, figure against ground, and freedom against its limits. These models seem all the more pertinent when we speak of mobile communication in its urban contexts. The weighty physical structures around which people and information move in cities serve to exaggerate the unmoored character of wireless communication. Mobile communication often may seem like a clandestine undermining of the city's structural solidity.

If the difference between variable events and stable structures defines wireless communications, that difference has often inspired an unease among artists and activists, as if it were the symptom of an unfinished revolution. Commercial and artistic experiments respond to this sense of incompletion by either destabilizing networks (endowing them with the variability of randomized human behaviours) or stabilizing human action (turning it into a predictable feature of network functioning). We find the first of these responses in certain works of locative media, like the *LOCA* project, which embed signal transmitters in obscure parts of the urban environment. When signal relays or transmitters are made to move in unannounced patterns, or to be triggered by unsuspected actions, we see the attempt to include network infrastructure among the variable and surprise-producing features of urban life (http://loca-lab.org).

Works such as these participate in the long-standing fantasy which casts urban space as the site of perpetual estrangement. The sensory-rich cities described by Charles Baudelaire and the Situationists presumed both the unconstrained freedom of the human stroller and the endlessly surprising, richly communicative character of the urban backdrop itself. As Thomas McDonough has suggested, the Situationists sought out urban phenomena which functioned, not as points of orientation or collective experience, but as the bases of a 'behavioural disorientation' which would produce the city as forever strange (McDonough 1996, 59). The revival of interest in these thinkers, over the past fifteen years, has coincided with the rise of digital wireless communications, influencing the terms in which technological experimentation involving mobile media has imagined and expressed itself. (A Google search of sites containing both 'Situationism' and 'wireless' produced 36,000 hits in January of 2010.)

Conversely, another set of innovations wants to make human action as solidly invariable as network infrastructures, turning human behaviour into one of the reliable supports of communications systems. Drew Hemment has written about Ester Polak's project *Amsterdam Realtime* (http://www.waag.org/project/realtime), in which participants meandered through a city with GPS devices that registered their movements on a public screen:

> At the outset the screen is blank, but as the journeys are recorded, individual meanderings fuse into a composite representation of how people occupy and use the city – density and concentration are recorded in the luminescence of overlapping lines; spaces unvisited remain dark. (Hemment 2006, 350)

The aggregation of variable movements into solidifying maps of predictable behaviour allows people to 'write' a new (and seemingly personalized) cartography of the city even as the same people become the enactors of collective patterns which, ultimately, become repetitive. The gap between the random variability of the participants' GPS transmissions and the network's drive towards cartographic stability is thus overcome with time. In his study of Parisian culture in the early twentieth century, Adam Rifkin describes the efforts by artists to combat the sameness which resulted when human movement within a city, however chaotic and subversively imagined, inevitably reproduced a stable complexity (Rifkin 1993, 157). For contemporary artists, the demonstration that individual behaviours aggregate within collective networks and movement systems is one way of revealing forms of community behaviour which may challenge official understandings of such behaviour. A project like *Amsterdam Realtime* offers, as one of its many 'outputs,' new knowledge about the city's unacknowledged use patterns – knowledge which may compel administrators to revise their understandings of how a city is inhabited and used. (This is one of the social benefits of the *Amsterdam Realtime* project discussed at http://www.interdisciplines.org/move/papers/3). For Rifkin's Surrealists, in contrast, the characteristic response was to embrace obscurity, to embark upon patterns of movement and behaviour which clouded any administrative understanding.

The transformation of variable behaviour into stable structures is evident in the more banal ways in which digital communication interfaces stabilize shifting relations of distance and mobility between people, disguising them with identificatory marks which betray little or nothing of human variability. The Skype interface, for example, announces incoming calls with the faces or names of the individuals making them, rather than with phone numbers, which (like postmarks in postal communications) would betray a caller's place of origin. Our contact lists register the slow growth of our interpersonal networks, rather than our constantly shifting relations of proximity to others.

Any theorization of urban communications must confront the relationship of stability to impermanence, of stasis to mobility. Versions of this duality are scattered throughout cultural theory's claims about cities and the communication which transpires within them. We may find this tension at the heart of Victor Hugo's oft-cited recognition that the printing press, as an agent of incessant disruption, would challenge the deeply anchored authority of the Church (Hugo 1831). It is not simply that one social force would, with time, eclipse another (as a result of democratization and secularization). The restless mobility of the press offered a more apt expression of the dynamism of collective urban life than did the architecturally entrenched power of the Church. More recently, Karlheinz Stierle has written compellingly of the ways in which urban languages flow promiscuously in speech and interpersonal exchange, then are captured, rendered immobile within enduring textual forms, such as magazines or commercial signage, 'just as cold comes to be fixed within ice' (Stierle 1993, 2001, 29, my translation). The historical work of David Henkin (1998), William R. Taylor (1991), and others on New York City traces the ongoing traffic of language between ephemeral acts of speech and the physical structures (such as billboards or newsstands) on which vernacular language comes to be affixed. (See also Darroch 2007 and Fritzsche 1996, for similar work on other cities.) Some of the most interesting theoretical thinking about the visual communications of cities has addressed the ways in which different media manifest the tension between the stable and the variable. Akbar Abbas, writing of Hong Kong, distinguishes between urban architecture and cinema, setting the stable spectacle of buildings against the partial and ever-changing viewing positions offered by films (1997, 64). In different contexts, Alain Mons (2002) and Pascal Pinck (2000) have written of urban photography and the 'skycams' of airborn news-gathering in terms of their constant stabilizing and destabilizing of the city as an object of sensory knowledge.

Distance Reading

The analysis of mobile communications (of cell phone conversations or SMS messages, for example) typically operates at either of two extremes. At the highest level of generality, we find infrastructure and system, studied from the perspectives of a political economy of media or as elements of urban infrastructure. At ground level, we find the flickering of impulses and signals, interesting for the micro geometries of human or social action which, through this flickering, are created or revealed. Mid-range phenomena, like the texture of messages and semantic substance of communicative textuality, easily drop out as significant concerns in the analysis of wireless devices. It seems much more pertinent, in the analysis of digital device communications, to pursue what literary theorist Franco Moretti has called a 'distance reading' – to undertake the analysis of small phenomena from afar, so that messages reveal little more than positions and linkages within social space. In a 'distance reading,' Moretti writes, 'distance is ... not an obstacle, but a specific form of knowledge: fewer elements, hence a sharper sense of their overall interconnection. Shapes, relations, structures. Forms. Models' (Moretti 2005, 1). What is sacrificed in a distant reading, of course, is the meaningful substance of communications. Social texture is no longer a world of meanings shared or contested, but a set of diagrams produced by the rhythms and geometries left as residues by successive communicational events. A distant reading seems appropriate to media forms (like the text-messaging mobile phone) whose textual expression is fleeting and, by most standards, trivial.

Perhaps because of this triviality, mobile device communications have received little attention within the rich renewal of media theory which has transpired over the last decade or so. This renewal, best characterized as a turn towards ideas of 'materiality,' is dispersed across several currents within cultural analysis. In contemporary humanities scholarship, for example, concepts long dormant or discredited within advanced media scholarship, such as 'transmission,' have been the focus of renewed interest and elaboration (e.g., Guillory 2004). This renewal has been most influential and provocative in the recent work of the so-called 'German school' of media theory, in the work of Friedrich Kittler, Bernard Siegert, Norbert Bolz, Hans Ulrich Gumbrecht, and others. Over the last decade, special issues of such journals as *New German Critique, Configurations, Critical Inquiry, Literary History, Theory, Culture and Society*, and the *Yale Journal of Criticism* have served to develop this ferment beyond this unitary German contribution, enlisting within it more localized developments in book history, architecture studies, film studies, and the social study of science.

This work is heterogeneous in its claims and points of departure, but united in its interest in the so-called 'externalities' of media forms. 'Externalities' does not simply designate those dimensions of media we might consider hardware or packaging. The term invites us to consider the materially embedded character of cultural expression, its inscription (as with writing) or iteration (as with performances) within arrangements of technologies, bodies, and physical structures. Media forms, this work argues, provide the contours in which cultural expression is contained and shaped; media forms store or transmit this expression in culturally pertinent ways. David Wellbery has usefully traced the theoretical genealogy of this interest, from Michel Foucault's imperative to 'think the outside' (Foucault and Blanchot 1987, 1–5) to the tendency of Kittler and others to think within a 'presupposition of exteriority' (Wellbery 1990, xii).

For all its emphasis on 'externalities,' however – on the defining contours of technological encasement and public interface – much of this work is concerned with the ways in which media gather up within themselves the communicative substance of urban life. Recent theoretical work has enumerated the characteristics of media forms in a list of functions whose degree of overlap is striking. Kittler's definition of media – as technologies for the 'storage, processing, and transmission of knowledge' (quoted in Griffin 1996, 710) – is echoed in other work on media or social discourse which may or may not follow his lead. Here, we find media forms defined by their capacities to 'absorb, transform and rediffuse' (Angenot 2004, 212), 'absorb, record and transform' (Heyer and Crowley 1995, xvi), 'circulate, record and organize' (Esposito 2004, 7), or to serve within processes of 'processing, storage and transmission' (Wellbery 1990, xiii) and 'preservation, transmission, and translation' (Knauer 2001, 434). Each term in these lists describes one way in which the media *work* upon the pre-given practices and semantic substance of the social world.

Unexpectedly, much of this recent media theory seems poorly suited for the newest of media technologies, such as digital hand-held devices. For all its attentiveness to computer technologies and scientistic knowledge systems, most of the work in media theory just described has found its fullest deployment in the study of print forms such as literature, wherein the sedimentation of social discourse is richest and most obvious. Mobile messaging may well 'absorb, record and transform,' in capturing bits of slangy speech or registering preoccupations (like ingroup gossip) which pre-exist and circulate through it. This capturing cannot help seem minor, however, set alongside the ways in which the literary novel or the theatrical performance absorbs social thematics and appears to organize knowledge and emotions. Likewise, the argument that literary forms are interfaces, or devices of storage and transmission, will seem innovative because these forms have rarely been thought of in these terms. These conceptions seem much more banal when used to discuss mobile communications devices, whose instruction booklets contain such terms in their description of functions. In the same way, the claim that scientific or artistic work may profitably be thought of in terms of network structures seems more genuinely novel than the claim that mobile communications might be considered in those terms as well. A variety of recent currents within media theory suffer, in their application to mobile device technologies, from being too close to the self-understanding of those engaged in their production and use.

The Circulatory Turn

Alongside this busy renewal of media theory, and interweaving with it at multiple points, we find a set of gestures within cultural analysis united by their concern with the idea of circulation. Here, the principal question concerning media is not their action in relation to some prior substance (like social discourse, knowledge, or subjectivity) to which they give form. Rather, the turn to circulation comes with an understanding of media as mobile forms circulating within social space. This 'circulatory turn' has been most influentially defined by Dilip Parameshwar Gaonkar and Elizabeth A. Povinelli as the imperative to study the 'edges of forms as they circulate' (2003, 391). Edges here are no longer just the outside, container elements of cultural substance. Edges constitute the interfaces of cultural artefacts with human beings and other forms, the surfaces which organize a form's mobility.

The nature of this intervention is an anti-interpretive one, intended to challenge a concern with cultural forms which sees them principally as bearers (however mobile) of meaning. The bases of this challenge to interpretation will vary across a range of recent interventions. For Hans-Ulrich Gumbrecht, the 'uncontested centrality of interpretation' within cultural analysis has blocked an attention to 'presence,' to the sensuality and tangible materiality of the cultural form (Gumbrecht 2004, 17). Put succinctly (even crudely), Gumbrecht suggests that the purpose in analysing cultural artefacts should cease being that of 'imagining what is going on in another person's psyche' (2004, xiv). Rather, it should address the conditions under which cultural forms occupy social space, interconnect and move in relation to each other. For Gaonkar and Povinelli, the most vital cultural analysis will be that which learns to 'foreground the social life of the form rather than reading social life *off* of it.' The key question is no longer that of how personal or collective life registers itself within communicative expression, but of how the movement of cultural forms presumes and creates the matrices of interconnection which produce social texture. In what might serve as an invigorating program for the analysis of mobile device technologies, they call for an ethnography of forms 'that can be carried out only within a set of circulatory fields populated by myriad forms, sometimes hierarchically arranged and laminated but mostly undulating as an ensemble, as a mélange, going about their daily reproductive labour of mediating psychosocial praxis' (Gaonkar and Povinelli 2003, 391–2).

Like the Kittlerian media theory described earlier, this intervention is more dramatic when it acts upon fields of inquiry long accustomed to unpacking the rich discursivity of a given artefact or practice. Of all the weaknesses that have plagued the analysis of wireless communications, however, a preoccupation with the in-depth interpretation of complex textualities has not been one of them. As suggested, the individual message, utterance, or text within wireless communications flickers in and out of the analytic gaze; that gaze, in turn, locks more easily onto a view of the overall infrastructure of interconnection. In our opening image of mobile communicators moving in multiple and uncoordinated directions, the rearrangement of bodily geometries will automatically seem more significant than the exchanges of words which reveal (or justify) such rearrangements. Even this scenario, however, still has human beings at its centre with mobile devices as the tools wielded in the maintenance of relationships across space. Theory need only take one more audacious step to see this scenario as one whose geometries are defined by the relationship between technological devices, rather than that between the human beings to whom these devices are attached.

'Circulation' was the theme of the April 2005 issue of the art magazine *Frieze*, an event which cemented the concept's move to the centre of art criticism's contemporary vocabularies. In the issue's programmatic statement on circulation, Jorg Heiser traced the concept's multiple roots. The work of Gaonkar and Povinelli, and of Benjamin Lee and Edward LiPuma (all of them writing in the journal *Public Culture*), was a key influence here. Such work, born within (or on the margins of) cultural anthropology, has come to the attention of an art world already engaged, for almost a decade, in working through the notion of a 'relational aesthetics.' Formulated first in the art criticism of the French curator/critic Nicolas Bourriaud, a 'relational aesthetics' offers the following premises as characteristic of a range of contemporary art practices:

> Each particular artwork is a proposal to live in a shared world, and the work of every artist is a bundle of relations with the world, giving rise to other relations, and so on and so forth, ad infinitum. (Bourriaud 2000, 21)

As Heiser suggests, the theoretical program for a relational aesthetics gestures towards 'circulation' without using the word. It does so through its emphasis on the moment of encounter, on the meeting of actors and works at determined places within social space. In Jacques Rancière's account, an aesthetics of the encounter assumes importance when it no longer seems radical, possible, or interesting to conceive the artwork as a critical engagement with the surrounding world of signs or commodities: 'Yesterday's distance towards commodities is now inverted to propose a new proximity between entities, the institution of new forms of social relations. Art no longer wants to respond to the excess of commodities and signs, but to a lack of connections' (Rancière 2006, 89).

At the same time, Heiser argues, a relational aesthetics, with its commitment to healing fractures in the social bond, bumps up against the limits of its own humanism. It remains too tightly moored to a vision of artworks as tools employed in the act of communication between people (between artists and their audiences):

> They [Boris Grouys and Nicolas Bourriaud, key theorists of relational aesthetics] consider the relationship between artists and the forms they propose (understood as the embodiments of ideas), and the interactions between people and the world triggered by it, but they seem to ignore the relation between these forms and other forms. The term 'circulation' is shorthand for the ways in which the fluctuating relations between forms (from both inside and outside art) co-define the relations between artists and their audience. (Heiser 2005, 79)

Heiser's most compelling phrase here – 'the fluctuating relations between forms' – quickly evokes the 'undulating' ensembles of forms imagined by Gaonkar and Povinelli. Both images are part of a broader imagining of communicational universes as populated by cultural forms talking to each other, their geometries of distance and interconnection made and remade in ongoing fashion. For Heiser and others, one advantage of 'circulation' is its displacement of 'production' and 'reception' from a cultural analysis which has spent too much time fretting over the relative primacy of each in the life of cultural artefacts. Circulation is not just a third level of analysis (like 'distribution' in the study of the cultural industries), but names the point at which production and reception have collapsed as meaningful moments.

Elsewhere, Michael Giesecke has used similar language to characterize literary production on the Internet:

> Pressured by new technological media, artists and scientists instantly orient themselves away from the notions of 'production' and 'reception' typical of industrial and print-based cultures. Literary projects on the Internet, among others, frequently do not allow for a strict distinction

between authors and users, participation and distance. (Giesecke 2002, 11)

We must be careful, however, not to confuse these arguments with those of Heiser or other theorists of circulation. For Giesecke, the distinction between production and reception disappears within a new participatory regime of creation which deepens the involvement of all concerned. (This is a common claim about Internet-based literary production, and was always already true of the telephone conversation.) To circulation theorists, the distinction loses interest as its constitutive terms themselves lose descriptive usefulness. The movement of a cultural form (a newspaper or SMS message) is not one which bridges a source and destination, but the realignment of forms in relationship to each other, within 'undulating ensembles' that give cultural life its character.

Circulation and the City

'In a given culture of circulation, it is more important to track the proliferating copresence of varied textual/cultural forms in all their mobility and mutability than to attempt a delineation of their fragile autonomy and specificity' (Gaonkar and Povinelli 2003, 193). As a slogan, this comes as close to any in capturing my vision of a cultural studies which is compelling. Those of us who study ephemeral forms, from SMS messages through old dance tracks, know that a preoccupation with content and interpretation will seem less powerful than an analysis of the ways in which these things are inscribed upon visible or sonic surfaces. Any one instance of SMS messaging or rhythmic sequence is less interesting than the re-mapping of the city which goes on as the edges of these things join together in series or pathways. To study the 'edges of forms' is to study not only the containers of meaning, but the systems of assembly and interconnection which give texture to urban cultural life.

Calls to study the 'materialities' of communication and the 'edges of form' represent significant interventions in media theory, but they invite us, as well, to revise our ways of thinking about the city. The city is constituted to a significant degree from these materialities and 'edges of form.' The routes traced by postal workers tell us little about the substance of letters or parcels, but in the organization of cities as systems of routes and addresses, a spatial rationality is built upon the circulation of intimate expression (see Siegert 1999). The physical edges of the city's built environment are almost inevitably mediatized, either through the functions they assume as points of orientation, or through the ways in which they become literal surfaces for inscription and text. Information or cultural expression does not simply blow weightlessly through the city, but becomes a pretext for the building of structures and the organization of space, for the fixing of interfaces (like the public telephone) to particular kinds of places and for the assembly of people around media nodes (such as the sports bar television) (McCarthy 2001).

Models of circulation will vary in the extent to which they emphasize the controlling character of circulatory processes or work to convey their open-ended flux. Clive Barnett has pointed to the divergent, even contradictory, meanings which 'circulation' has assumed within cultural analysis. On the one hand, the word may designate a 'circular, tightly bound process,' the setting in place of control systems through repeated patterns of movement and the building of stable structures to channel this movement. This sense of circulation is powerfully conveyed by Kaika and Swyngedouw (2000) in their account of the development of urban infrastructures for transporting water, electricity, and gas. In this historical analysis, the development of cities has involved the ongoing integration of natural forces within technologically based circulatory systems. On the other hand, 'circulation' can suggest a 'scattering and dispersal,' the dissolution of structure within randomness and uncontrolled flux (Barnett 2005). The movement of news, gossip, money, commodities, and people, outwards from their places of origin or routinized departure, involves the ongoing drawing and redrawing of circulatory maps whose variability works against any sense of stable urban infrastructure.

The difference between circulatory processes which are 'tightly bound' and those marked by 'scattering and dispersal' returns us to the tension with which we began. Cartographic mapping processes which register individual movement through the city capture the scattering of human action even as repetition deeply etches these passages within regularized, 'tightly bound' patterns. We are used to thinking of cities in terms of speed and mobility, but the complex circuitousness of cities also makes them places of delay and blockage, of a tightening. A city's circuitousness means that movement within it is stretched out, that things and practices and connections are given more space and time in which to eventually disappear. We might ask, in studying wireless device communication, what, amidst the novelty and rootless mobility of such communication, is being perpetuated: the friendships which would otherwise fade away were there not this low-commitment means of staying in touch? The connections to an earlier home or occupation or interest, which are now readily available for resuscitation?

A commitment to circulation as the orientation of analysis does not presume the claim that life is more mobile, fleeting, or fragile in the world of new media technologies. It does require attentiveness to the ways in which media forms work to produce particular tensions between stasis or mobility. When wireless computing, in and around hotspots, builds on much longer traditions of café-based observation, contemplation, or reflection, mobilizing the same sorts of literary ambitions which motivated Baudelaire or Edgar Allan Poe, it is serving an inertial function, slowing down the disappearance of ideas and sensibilities from the world. This sense of wireless computing challenges the more conventional understanding which casts it as the precondition for endless mobility. Discussions of technology are forever balancing competing images of a heightened mobility and a demobilizing entrapment. Diagnoses of an impending obesity epidemic mobilize the spectre of the immobile computer user just as advertisements emphasize the fluid groundlessness made possible by gadgets. Everyday complaints about new communications devices are as likely to emphasize their capacity to immobilize us, within the traps of responsibility and availability, as they are to trumpet their capacities for freedom and escape.

3 Radio Hats, Wireless Rats, and Flying Families

KIM SAWCHUK

In *Virtual Geography: Living with Global Media Events*, Mackenzie Wark comments that 'we no longer have roots, we have aerials' (1994, 7). For Wark, this aphorism signals the arrival of a postmodern state of perpetual, high-speed mobility within a world that is shrinking, accelerating, and imploding because of the proliferation of telecommunications. Wark is not alone in his depiction of the increasing dematerialization of social relations and people from a sense of rootedness in place (Baudrillard 1983, Castells et al. 2007).

Life in a world of instantaneous communication, media saturation, and accelerated mobility is a matter of academic and corporate concern. In addressing consumers, cellular telephone companies literally capitalize on the dreams and anxieties unleashed by the very technologies they sell. For example, in a series of Canadian advertisements for Bell High Speed Internet, the chubby buffoon-like beavers, Frank and Gordon, are shown at a busy intersection negotiating traffic, which threatens to turn them into road-kill. Telus Mobility advertisements for digital cell-phone services routinely depict cute animals who assure us that 'the future is friendly.' Paradoxically, that we must be assured the future is friendly, and not hostile, is surely a sign that the opposite is possible.

While we are in the midst of new corporate mergers, shifts in the delivery of media content, and a rapidly expanding diffusion of cellular services worldwide (Townsend 2004), the fascination with wireless communications and mobility is not wholly synonymous with our present moment. At the turn of the century, Nikola Tesla proposed the possibility of the 'World Wireless' (1915, 1919). In the 1930s, announcements such as 'The Cable Photo Is Perfected' (1938) and of 'Radio Pictures' (1939) describe the world to come with wonderment.

The corporate cross-media promotion of wireless technologies pre-dates the actual appearance of the cellular telephone in the hands of businessmen in the 1970s. For example, in 1944, Motorola's 'handie-talkie' was advertised with the Warner Brothers film *Objective Burma* (with Errol Flynn). In their promotional campaign, the company boasted: 'The Motorola Radio "handie talkie" plays a vital role in this great picture as it has in every battle front in this war.' Accompanying the advertisement for the handie-talkie, and the film, is a plug for the Motorola 'playmate,' which is an AC/DC powered portable radio. Prefigured in the ad is how research and development into wireless during the war will benefit consumers after the war.

While the world of wireless communications emerges in the landscape of corporate advertising before the war, with the end of the Second World War in 1945 the promise of a wireless world of unfettered mobility, asso-ciated with radio, television, the telephone, and the telegraph, takes flight. General Electric, RCA, Motorola, Bell, and Philco are but a few of the companies investing in the technologies and services that form its technical and material infrastructure. In this period of intense advertising following wartime research and development, a fascination with the power of 'electronic energy' merges with fantasies of the propulsion of the body into space with ever greater speed to form the basis of what I am calling a stationary mobility and the wireless imaginary.

These renderings of the wireless imaginary promote new forms of trav-el for white middle-class American families, who need never leave the safety of the home environment but who, nevertheless, can partake in events elsewhere at any time.[1] In this chapter, I trace this emergent wire-less imaginary, and the idea of stationary mobility, between 1945 and 1950 in that most iconic of photojournalistic magazines, *LIFE.* As Erica Doss (2001) writes, '*LIFE* played a major role in representing and dis-seminating information, ideas, and shaping their meaning to an ever in-creasing body of consumers fluent in the language of pictorial communi-cation' (4). It is thus a fitting locus for a journey into this still present past.

The notion of a collective imaginary circulates within multiple sites of theoretical inquiry, most of them deriving from more sophisticated ren-derings of traditional Marxist theories of ideology. Carlos Castoriadis (1998) theorized the social imaginary in *The Imaginary Institution of Society* to explain the social intertwining of the individual and collective psyche, actualized in social praxis (7). Documenting the collective im-aginary within modernity has been picked up in a number of recent works, including those of Charles Taylor (2002), who argues that a social imaginary is a set of stories and narratives distinct from a set of intellectual schemes (106).

While distinct, these theoretical perspectives on the imaginary share a commitment to understanding socially shared representations as they are connected to the construction and creation of institutions, such as democracy, and the practices that sustain them. As Taylor eloquently writes: 'This understanding is both factual and "normative"; that is, we have a sense of how things usually go, but this is interwoven with an idea of how they ought to go, of what missteps would invalidate the practice' (106). What gets set into place is a kind of 'background understanding' (107).

What follows is a reflection on stationary mobility and the emergent post-war wireless imaginary, whose protean shape swirls around three clusters of images-texts that demarcate the 'structure of feeling' of this period. First, the images convey a fascination with electrical energy and the idea that with new systems of networking, words can fly. Second, we see the promotion of an imagined – and newly reconstituted – white, middle-class urban and urbane post-war family taking stationary journeys. Third, with the articulation of advanced telecommunications systems flying through the air, a new form of perceptual apprehension, 'aerial perception,' is instantiated, rendering a worldly sensibility that national borders and boundaries are potentially breached, and thus in need of protection from 'others.'

In examining this cluster of images, one sees how the malleable shape of this wireless imaginary was inscribed in and promoted within the mass media and popular culture. The contours of this imaginary are not sedimented into immutable discourses. Rather, drawing upon Raymond Williams's (1977) evocative oxymoron 'the structure of feeling,' these social and technological forms are active and flexible, not fixed and finite (128–9). Writing on literary forms, Williams uses the term 'structure of feeling' to express 'a particular quality of social experience and relationship, historically distinct from other particular qualities, which gives a sense of a generation or a period' (131). In the period following the Second World War, a fascination with the power of electronic energy merges with fantasies of the propulsion of the body into space with ever greater speed, promoting new vistas and forms of imaginary travel for families.

Electrical Impulses and Words with Wings

Electronics, electromagnetism, and electricity are central to this reflection on our dreams of a wireless mobile world. Electricity, as Carolyn

Marvin (1990) so aptly documents, is integral to the history of communications, nation-building, and modernity. In his discussion of the transmission model of communications, and the history of the telegraph, James Carey (1989) argues that this history of electricity and the telegraph may be secular, but that it has deeply religious origins. This secular religiosity is evident in the images depicting the very process of transmission, such as a famous 1861 allegorical drawing depicting American telegraphy (Schivelbusch 1979). In this image, an angel carries a rolled tube of parchment with the message 'May the Union be Perpetual' inscribed on it. While the angel has wings, it does not fly. Instead, it trots along the wire, message in hand. In this image of telecommunications, the power of electricity is associated with the heavens, as lightning and angels are associated with the sky. Electricity, often depicted as a bolt of lightning, had given words wings, an image conjuring the invocation of the gods. This is a powerful trope that returns in the promotion of electricity during and after the war.

With the unfolding of the 1920s and 1930s, the term 'wireless' was associated with radio transmission. Television, the scanning and transmission of pictures with electrons, was introduced at the 1938 World's Fair along with radiogrammes. Sending pictures electronically was also newsworthy to *LIFE* editors in 1939. But these experiments in early analog wireless broadcasting were just that: experiments largely interrupted by the next war. What all of these consumer items depended on, of course, was electricity in the home. Electricity, by no means in every North American home in the 1940s, particularly in rural areas, was a subject for magazines like *LIFE* and their powerful corporate advertisers. General Electric and the Electric Light and Power Company were at pains to remind consumers during this period that electricity was key to modern living, and to the movement from an agricultural society to city living.

In a full-page advertisement from the Electric Light and Power Company in 1945 – 'Why did Aunt Hallie wrap the ice in paper?' – a young man questions an elderly woman on her antiquated methods of preservation during the 'the olden days' before the refrigerator. Electricity saves money, cuts down on unnecessary labour, and is the modern way to preserve food from spoilage. It is associated with youth and the future. In another ad during the war, a superhero with wings on his head and bolt from the sky stands above a home, 'electrifying it.' The ad copy tells the story of the bright new world soon to come, once the hardship of war is over: 'Now M-H Electronics are at war, performing miracles in planes, tanks and ships.' However, the promise is that these 'same miracles, or modifications of them,' will be made available 'to homeowners everywhere after peace comes.' Just what miracles would be conjured were spelled out for future consumers in these advertisements. Electric Light and Power from this period, for example, depicted six functions of electricity for modern living, which included X-rays, telephones, concerts in the home, and spectator sports at night. In the advertisements during the war, electricity was a code word for modernity, progress, and the future to come after the sacrifices entailed by war were over.

In the visual landscape of the 1940s, several visual conventions and strategies crystallize to convey the power of electrons and electricity in easily recognizable graphic form. An identifiable iconography emerges, a visual shorthand that is distinct to the technology and the period. Two are most prominent: the visual tropes of the stylized lightning bolt and the Rutherford model of the atom, both used in this period to signify the power and might of electrical energy. The implied power of electrical energy is identifiable in an instant in the shape of swirling electrons or a jagged line, making the invisible visible through the power of a visual metaphor. These icons draw upon ancient mythologies: the M-H superhero bears an uncanny resemblance to Hermes, the Olympian god of travel, literature, and commerce. Hermes and Mercury, his Roman counterpart, are typically depicted wearing a traveller's hat with wings, or winged sandals. Hermes is most commonly noted as a messenger from the gods to humans.

In the post-war period, companies would make ready with the promises presaged in the advertisements during the war, transforming research and development in 'electronics' used for military telecommunications into consumer goods. The application of radar technologies, developed during the war, would lead to investments in wireless message systems to send messages soaring through the air on a solo flight. Particular media technologies, such as portable radios and television, were sold on the pages of LIFE. Also on offer was the very idea of wireless communications, often phrased in ad copy as 'the journey of a word' or 'words with wings.' In telegraphy, one had evidence of the materiality of the communications process in the form of a wire running from pole to pole. These ads liberate words from the wire.

Anticipating the concerns of the incredulous that messages could be lost in space, or disappear into the ether, some corporations went so far as to depict how invisible waves could reliably transmit a signal.

Depicting the movement of a signal or impulse as a discernible visible entity was achieved in imaginative ways in the post-war period that cashed in on the *caché* of these technological achievements as an aeronautical event.

Take, for example, Western Union Telegram's advertisement for services that appeared in *LIFE* in 1947. 'Radio wings speed along 2000 telegrams at once,' reads the headline, while the caption adds, 'by ultra-short radio waves.' In the image, heroic little 'radiogrammes' in the shape of letters move through the air assembly-line fashion, and do so without the necessity of human presence. Depicted in the ad are towers and radar dishes (at a distance of fifty miles each) to indicate that the messages will be carried along a broken line of electrical energy that will dependably take the place of the telegraph wires. While invisible, advertisements in the period make the case that this form of transmission is even more dependable than the previous mode. Not only can more messages be carried, simultaneously, but these messages moving reliably along these man-made bolts of lightning will be unimpeded by the terrestrial forces of geography, without battling the electrical interferences produced by nature in the form of weather systems like lightning or storms.

What is being sold here is not a machine or a piece of hardware but the affective qualities associated with the technologies: reliability, speed, and capabilities of a mode of transmission that was not immediately visible. What must be visualized is the very idea of a 'network,' a new word for what another ad called the new 'chain of communications' transforming the North American landscape. This early conception of a wireless networked space of antennae and towers is an assemblage of relays and switches from hill to hill. Like an intricate game of connect the dots, this assemblage collapses the perceived distances between major cities that were at that time being joined across America, and around the globe.

In depicting the revolutionary means of the new electronic technology, the older means of transmission and communication, already known and familiar, are referenced. As in the introduction of many new technologies, these first attempts to create a common vocabulary draw upon the dominant wireless media of the time, in this case radio. This McLuhanesque rear-view-mirror look at media is used to sell the virtues of the new with reference to the familiar, to instill trust in the unknown so that its services will be purchased or used. The main target audience, the ideal consumer configuration and the key to post-war prosperity and consumption, at least in a magazine like *LIFE*, is the white middle-class family. The history of mobile technology is, as Judith Nicholson points out, profoundly racialized by the constant and consistent inclusion of rosy-cheeked middle Americans, its promotion of narratives of progress spearheaded by a bevy of corporate males, and its exclusion of seemingly peripheral tales of how these technologies became central to the war on crime, drugs, and now terror (Nicholson 2008).

In the Family Circle: Stationary Mobility

Central to the post-war period are concerns about the fate of the white middle-class family. On the one hand, the family was depicted as being threatened by high divorce rates, too few children (1947), and what Lynn Spigel (1992) has called a discourse of 'Momism,' in which women who were at home during the war without the countervailing fatherly force were thought to have too much power. Most of the images of the family in *LIFE* depict a family unit of two or three children, the eldest a young boy, as the best configuration for reinstating the normal order of healthy gender relations after the disruption of war took father from home to the front, and made mother the head of the household.

The new electronic media of radio and television, and the new communications services like the telegraph and telephone, are regularly featured as central to family occasions. Advertisements for RCA Victor depicted this idealized family reunited in a variety of ways through technologies. And here it is important to note the arrangement of bodies and technologies. Family members were commonly depicted in a contained circle, a pictorial convention from the Victorian era, while at the centre of the image, and the clear focus of attention, is the technology itself.

These new post-war radiophonic technologies, such as FM (frequency modulation) as a means of transmission, also created unique possibilities for reconstituting the family. In one particularly fetching rendition of these RCA advertisements, the family unit of Mom and two kids resides inside the picture tube (RCA 1947e). While Dad appears to be strangely absent, a small white male figure is present as a scientist at the bottom of the page. What is depicted here is a trajectory from war to peace, a transition from the battlefield to the home, and now in the post-war period, from workplace to home.

This dynamic is captured in a 1944 Philco advertisement that boldly declares: 'FROM RADAR RESEARCH TO RADIO FOR YOUR HOME.'

While researchers had been busy inventing devices to be used in battle, these would now be used in peacetime: 'in this brilliant record of war research lies your assurance for the future ... when the Philco laboratories turn from radar to radio for your home.' The messages are unswervingly reassuring. In advertisements from this period, this narrative promise is repeated over and over. These goods are indications of the benefits of transferring scientific and technical knowledge to the domestic environment as the spoils of war.

In striking contrast to advertising images in our own period, the graphic style of these advertisements is unique to the period: a large central image, a short bit of ad copy at the top, a long written text, a small inset in one of the corners. In terms of graphic design, these ads were typically grounded by a small insert at the bottom or the top of the page, which would show European and American scientists hard at work in a real laboratory (or the laboratory itself). Another caption would situate this small image and explain the connection of one space (the world of technical production) with the home. The central fantastical dream imagery and slogan, associated with a domestic environment, is grounded in the 'real' world of scientific enterprise, which is imagined as bringing about the technological innovations to the post-war consumer, who is unswervingly depicted in racialized terms as Caucasian.

These inserts often explain in length and in detail the intended meaning of the more imaginative image and the technical achievement that has made the product possible, calling upon the authority of the scientific and industrial basis of the technology being promoted. These fantastical images were articulated to a real place or person, usually named to convey the authenticity of the message and to show that the 'wonders' on offer were not a pipe dream, but achievable through science and technology. Yet in this division of the phantasmatic space of consumption, contrasted with the 'real' world of production, a racialized gendering of place and space is set into motion: the feminized domestic environment is contrasted with the world of work inhabited by male scientists.

The idea that *worlds* are opening up after having been closed down, prohibited, or denied during the war is emphasized repeatedly. In another image from a now defunct television company, Stewart-Warner, the family is shown seated in front of a small screen, the world opened up to them. In another RCA ad, another family watches the radio whilst a world of music emanates from a cloud behind them. Images such as these underscore the potential uses of the technology in re-establishing the rhythms, patterns, beats, pulsations, of what was being promoted

as normal daily life. As Lynn Spigel's (1988) research suggests, the installation of the television played a crucial role in structuring the space
of the post-war home and reflected and shaped family relations. What
is intriguing in this set of advertisements is the manner in which the
introduction of these communication technologies was also seen as a
source of adventure for imagined travels beyond the home for particular types of families. This does not negate Spigel's claims. This contradictory tendency of simultaneously fantasizing about greater mobility
and validating the enclave of *this* home became part and parcel of the
post-war period.

Promoted in this period is a sensibility that I call stationary mobility,
through which one can travel without actually moving. One can venture out into the world, ensconced in the safety of a place that is familiar. One is simultaneously encouraged to inhabit two locations at once,
to be here and there. This is made possible through the proliferation of
media technologies that augment and extend the sensorial experience
of moving without going anywhere.

References to travel abound in this period, both virtual and actual.
Ad copy for radios, such as the 'Trav-ler,' invited listeners to 'listen …
and be carried away! For Trav-ler has achieved an amazing tone real-
ism that makes listening like traveling' (1947). To underscore this form
of imaginative journeying from the comfort of one's own living room,
the ad shows a man leaning back in a comfy chair smoking a pipe and
looking dreamily into space in the bottom of the image. In the opposite
corner, at the top, is a drawing of a voluptuous Spanish dancer twirling
under a tree. Indeed, the figure of a woman, coded as the racialized
exotic other, serves as the imagined destination and dream of desire in
the image, a narrative phenomenon analysed by de Lauretis in the
pages of *Alice Doesn't.*

Travel was highly coded by the inextricable weaving of gender, race,
and class privilege as the above advertisement indicates. Stationary
journeys could be taken by families, and destinations included the imaginary play-space marked by orientalist fantasies. In an astonishing
advertisement for General Electric, consumers are promised that 'more
and more people will go sightseeing as the number of stations and
home receivers increases' (1948). In a sanitized appropriation of Arabic
culture, we see a gravity-defying family flying together on a magic carpet ride. Dad sits at the helm pointing to what's ahead, mother rests by
his side, sister hangs on smiling, and brother (already a little man) seriously looks on. The nose of the carpet, one soon notices, is in the form

of a plane, which is itself the architectural shape of the RCA plant in Camden, New Jersey. It is a remarkable set of associations in which architecture, airplane design, and tourism are condensed into one image. In this promotion of stationary travel, by way of the electronic media, the depiction of a world in movement is full of the promise of adventure. During this period, tourism and travel were being heavily promoted, and real American families (read patriarchal, white, and middle-class) would be offered alternative ways of vacationing without ever leaving the comfort of their secure domestic environments. This discourse, however, reveals silences and borders: women were never depicted alone, or in movement beyond their household, without their family, or their husbands, chaperoning them. There is, as well, what Nicholson, following Stuart Hall, calls an inferential racism at work, a racism that operates by what it does not need to say but instead provides the foundation for the assumptions of what is considered 'normal' (Hall 1990).

These advertisements indicate how communications technology could act as a powerful 'technology of gender,' to quote Teresa de Lauretis (1987, 15), making the family an even more cohesive unit. But as well, what they act as, is a *technology of race* that consolidates the inferred assumption of the time: that African-American families are only dysfunctional, and that their neighbourhoods are inherently insecure and a hotbed of potential criminality (Nicholson 2008). This is evident in a *LIFE* article about the erosion of 'family values' and the decline of Western civilization. In short, the imagined family from the 1940s was bundled together as a *unit* safely ensconced within the locus of the home. This home was being modernized and undergoing important transformations in the post-war period, with the move to pre-fabricated homes, and an impetus to institute new regimes of hygiene and germ-free living that stressed a new dawn of cleanliness and togetherness (Wright 1983). If life during the war had been a time of rationing, hardship, and personal sacrifice, then in this post-war image of modern living with telecommunications, things were looking up.

Reaching Out, Reaching In: Aerial Perception

I mean this figuratively, and literally. There is a fascination in this period with the air: airplanes, radar, and antennae. Everyone, everything, everybody, could, potentially, be outfitted with an aerial to extend their range of personal reception. Aerials and the installation of radio towers were newsworthy stories in *LIFE*, such as the 1949 story of shoppers trying on the 'The Man from Mars Hat,' which promised users fantastic radio reception and new mobility for ambulatory listeners within a twenty-mile range of any radio tower.

While there had been experimentation with using the ionosphere and the moon as a way to bounce radio waves, many of the rocket launches that attempted to control these ventures in wireless transmission had been failures (Parks 2005). Aerials, whether on the head or on a hill, were the architectural signifiers of this new mobile connectivity through wireless, albeit one-way, transmission. The tower and the antenna were the obelisks of the 1940s, tied as they were to the new media of radio and television, the construction of man-made landscapes, and the new architecture of cities then springing up in the form of skyscrapers.

What was needed were ways for the body to reach higher altitudes from earth. If you could not have a personal plane, well, you could at least use Texaco sky chief. In this famous ad campaign, cars were shown soaring through the air. Associated with early forms of mobile, wireless communications, and the idea of stationary mobility, is a mode of perceptual apprehension that can be termed 'aerial perception.' Perceptual apprehension signals the ways that we haptically and sensorially grasp the world cognitively through the body in movement (Parviainen 2002). It also references the visceral anxieties that may be induced when new forms of movement are introduced into a particular cultural context. Aerial perception, associated with the airplane and the rocket ship, can be contrasted to what Wolfgang Schivelbusch (1979) calls 'panoramic perception,' which was introduced with the transition from passage along roads by foot and then coach to travel by train in the late 1800s.

In *The Railway Journey*, Schivelbusch documents the corporeal effects of the mechanical motions of the train, which replaced horse and buggy. This new mode of transport brought new velocities and senses of distance to travel. It made the foreground evanesce, but also made those unused to it quite literally unable to see out the window, as their vision was unaccustomed to processing the stream of images passing before them. Looking out a window, at the panorama receding in front of one, transformed the landscape. Systems of tunnels were built through mountains. The tracks themselves were accompanied by telegraph poles and ribbons of wires carrying information from place to place (66–7).

Aerial perception transforms vision and one's relationship to space in another manner, by providing a view from the air (Virilio 1994). In *LIFE*, this new form of perception is documented in pictures. In one image, we see the city of Manhattan from above. The tops of the roofs form a set of small square patterns on the page because one is looking directly from above. Images begin to appear taken from satellites and rockets. These images from unattainable heights gave the public a perspective from the skies, creating new cartographies of countryside and city. If one could not afford to take an airplane trip, these journeys into space could be experienced by simply reading the magazine and looking at the images.

In advertisements for new forms of travel and stationary mobility, these technological developments were articulated temporally, and without shame, with respect to other events in recent history. The Allied victory during the Second World War, and the effects this war had, or was thought to have had, on 'normal' family life would be visually invoked time and again. Radar, in particular, was celebrated as the technology that won the Battle of Britain, and this was capitalized on when selling to businesses who were engaged in their own little battles for strategic advantage against any of their competitors. A secret wartime weapon, radar, would acquire new uses in the post-war period: it would, for example, make air travel safer, claimed Howard Hughes, who was the first to put it on all TWA airplanes. It would be celebrated in popular and consumer culture. However, the language of consumerism would not displace the language of war or a discourse of external threat. The spectre of a new war, this time against a communist threat, would be mobilized to connect past traumas with future threats that would require constant vigilance.

While these new views of the world from distal heights were perhaps thrilling, they were also accompanied by a promotion of risk and danger. In another RCA Victor ad, 'Radio eyes,' we see an airplane itself wearing a pair of radar screens shaped as a pair of spectacles. Radar made use of a wider spectrum of communications and deployed frequencies that were previously undetectable for 'blind flying,' that is, flying without actually seeing ahead. The promise was that passengers would have greater safety, for with radio eyes one could see through and beyond whatever darkness or impediments nature might present. A subtext in all of these promotional images is that nature poses dangers and that the topography of the earth, with its hills, valleys, and other topographical irregularities, was a nuisance that could be overcome by technology. Aerial perception opened up and required new technologies of vision. If greater speeds and heights had been attained through aeronautic innovation, then radar was posited as a panacea to quell anxieties about these new forms of travel into space. Technological instruments would become the means to guide movements at these new heights and velocities safely.

In contemplating these images, a series of paradoxes, what Sara Kember calls 'fault lines' (1998, 87), open up. On the one hand, the discourse of telecommunications and the emergent wireless imaginary, and the discourse of stationary mobility, promise to create one common world, unite white American middle-class families, and bring modernity and progress. On the other hand, supersonic velocities of movement through space, which collapsed distance, also instilled a heightened sense of vulnerability and threat on a new worldwide scale to these very same family units, from 'motion sickness' to threats of invasion by powerful external forces. Speed and heightened mobility were not only providing new adventures. Their very existence meant the necessity for vigilance against threats from foreign bodies who might invade and infect the body politic. The ability to 'reach out to the world,' the slogan for Bell Telephones in this period, is telling. Such a slogan and the advertising image that accompanies it depict the new lines of global movement. However, what could go out could also reach in. Take, for instance, a series of ads in 1949 for the 'shell with the radio brain.' Consumers are at once being told how it could be used to shell enemies from a distance, yet it is also implied that if they are within our grasp, then we are within theirs.

In other words, the selling of these new technologies continued to promote the idea of war and risk, to install the idea of the need for permanent war and continued vigilance as the world was brought closer. This would set the stage for the impending Cold War, with fears of atomic attack and communist invasion looming on the horizon. These threats came from both without and within, from cosmic rays to the world of previously invisible germs (Ostherr 2005). New forms of perceptual extension underscored new risks and dangers, enemies lurking underneath the surface of things, above and below: from Russia, from China, from North Korea. Indeed, it is worth recalling that this is not only the period of the Cold War, but also the period of intense postcolonial struggle for national liberation in the aftermath of war throughout the globe. Enemies were everywhere and needed to be brought into visibility: they were in the air, in our bodies, in our minds, in other countries. They would need constant monitoring.

The language of constant warfare, even in times of peace, was used to give authority and credibility to the scientific endeavours that had been appropriated for wartime use. It was also mobilized to make consumers in the period anxious that if we do not keep up, we will be left behind in whatever race is out there. If shells could be given 'radio brains,' could our brains be wired by radios and controlled remotely, without wires?

Experiments with radio-controlled rats – the rats with antennae implanted in the tops of their heads – were recounted in one story of laboratory experiments at UCLA attempting to test the possibility of the wireless control of the species. This fear of remote control and brainwashing gave rise to a plethora of films in the 1950s, including *The Manchurian Candidate* and *Invasion of the Body Snatchers*, warning of the possibility of direct control over the thoughts and minds of others. Such imagery fed into an already existing discourse of hypnosis, propaganda, and electromagnetism. These depictions underscored the need for an 'arsenal' of strategies and technologies, including radar, experiments in behavioural science, extensive governmental funding, and research contracts to private industries, to keep potential enemies at bay. New technologies of detection would reveal new enemies, thereby justifying these investments and expenditures. Radar was not only a way to make civilian air travel safer, but also a means to reach into the sky, to put the finger on the enemy who might come from there, infiltrating our cities, invading the body, taking over the mind.

Conclusion

Flipping through the pages of *LIFE* from the 1930s to the present, one notes the emergence of a discourse that resonates in the present. Keywords like 'portability,' 'mobility,' and 'speed' begin to be used by companies such as GE, Motorola, Electric Light and Power Company, Western Union Telegraph, Philco, and Bell to sell an astonishing array of goods, from radios to aspirin to airline travel. Stories of miniature cameras, Dick Tracy radio transmitters, 'radio hats,' and rats with aerials attached to their heads begin to appear. These evocative texts and images, often taking up a full page in the magazine, tell the tale of exciting new vistas, new freedoms, and new means of transport without ever leaving home.

These corporate images are a part of an intertwined emergent sensibility of stationary mobility and the wireless imaginary that marks one dimension of the structure of feeling in the post-war period in North America. These moments, which exist at the periphery of official histories of wireless mobile technologies, mark the conceptual edges of this emergent sensibility as deeply connected to real events. The mediated appearance of these technologies are tied to imagined journeys and modalities of travel that paradoxically sediment familial power relations, thought to be under attack after the war, while also feeding into the creation and revelation of new enemies. What is revealing in these images is not only what they depict, but what they exclude. What is never imagined is that the need for security, or the desire for mobility, could extend beyond a very contained notion of family: Caucasian, middle-class, and decidedly urban. In this way, these forms of imagining communications, media, and movement act as a technology of gender and race.

Sarah Kember (1998) writes that 'one of the things that the past reveals is that many of the reactions to new technologies are not in themselves new' (2). She counsels historians of media images to embrace 'fantasmatic thought' to reveal the instability of social fault lines of power and desire (87). In this emergent culture, the hope for the future and a belief in its promise were underscored by fears of its fragility and possible collapse into a chaos that would undermine 'civilization,' a code word for all that is the legacy of a European ancestry. Media historian Caroline Marvin (1990) reminds her readers that 'we are not the first generation to wonder at the rapid and extraordinary shifts in the dimension of the world and the human relationships it contains as a result of new forms of communication, or to be surprised by the changes those shifts occasion in the regular pattern in our lives. If our own experience is unique in detail, its structure is characteristically modern' (3).

The stories and advertising one finds in LIFE in this period depict the achievements of science and technology in stories connected to past and future wars. By highlighting the forgotten anomalies, the strange experiments, para-events, and imagery that accompanied the creation of the first public radio towers to transmit 'words with wings' in the 1940s, we ourselves travel into the murky world of the past, which, as historian David Lowenthal (1985) writes, 'is a foreign country.' This wireless past, this distant country, does not provide answers once found or revisited. What it does provide is a back-story to our present condition, which may allow for a richer understanding and a perspective for future ruminations on the uniqueness – or not – of our current digital wireless mobile world.

Tracing these moments, which exist at the periphery of official histories of wireless mobile telecommunications, allows the feminist researcher to explore the conceptual edges of an emergent wireless imaginary tied to stationary mobility that, paradoxically, sediment familial power relations. Such tales encourage researchers to pay attention to the private and public politics embedded in these strangely unsettling, but uncannily resonant, phantasmagoric limits and moments.

44   Kim Sawchuk

ACKNOWLEDGMENTS

I would like to thank the staff at the Fraser Hickson Library in Montreal for their assistance in conducting this research. Ongoing dialogue with my friend and colleague Barbara Crow has helped to shape these ideas. Judith Nicholson has brought new insight to this project through her prescient analysis of how mobile technologies are connected to a culture saturated by hierarchies of race and power.

NOTES

1 There is, of course, as feminist geographers (G. Rose 1993) and philosophers of nature (Merchant 1983) have pointed out, a whole assemblage of historically gendered assumptions at play in these seemingly benign images.

ARTICLES

'The Cabled Photograph Is Perfected: The Picturegram.' LIFE, 14 Sept., 1938, 10.
'Radio Pictures.' LIFE, 13 Feb. 1939, 22.
'Motion Sickness: Willing Students Take Rough Rides in Experiment to Find Out What Makes Travelers Get Nauseated.' LIFE, 12 Feb. 1947, 54.
'The Family: In Western Civilization It Is Seriously Threatened and Needs Material and Moral Help.' LIFE, 24 March 1947, 36.
'New York Skyline Photo' (picture of the week). LIFE, 31 March 1947.
'Tourists: Man-Made Spectacles Rival Scenery as an Attraction.' LIFE, 7 April 1947, 134.
'Miniature Wrist Radio: US Bureau of Standards Develops a Dick Tracy Wrist Transmitter Which Can Broadcast Messages for a Mile.' LIFE, 6 Oct. 1947, 61.
'Boxes for Baby: New Style Crib Eliminates Germs, Drafts and Constricting Clothes.' LIFE, 3 Nov. 1947, 73.
C.J.V. Murphy. 'The Last 500 Feet: With New Inventions the Airman Is Piercing the Final Weather Barrier.' LIFE, 12 Dec 1947, 82.
'Tiny Camera: They Are Becoming Popular Gadgets for Kids, Sportmen and Snoopers.' LIFE, 20 Jan. 1949, 51–2.
'Radio Hat: Two Tube Set Plays Fine but Looks Ridiculous.' LIFE, 6 June 1949, 20.

ADVERTISEMENTS

American Railroads. 1947. 'Round Trip to the Moon.' LIFE, 28 April, 9.
Bell Telephone Systems. 1948. 'Journey of a Word.' LIFE.

– 1949. 'The Worlds in Reach Again.' *LIFE*.

Electric Light and Power Company. 1945. 'Why Did Aunt Hallie Wrap the Ice in Paper?' *LIFE*, 15 June, 25.

– 1947. 'How Many of These Need Electricity?' *LIFE*, 6 Oct.

Every-ready. 1949. Every-Ready Mini-Max: The Shell with the Radio Brain: Army/Navy Lift Censorship on Mystery Weapon That Licked V-Bomb, Kamikaze Attacks.' *LIFE*.

General Electric. 1948. 'Electronics Park.' *LIFE*, 19 Jan.

Gillette Blue Blades. 1947. 'Speaking of Speed, Chalmers "Slick" Goodlin Uses Gillette Blue Blades.' *LIFE*, 2 Jan., 32.

Motorola. 1944. 'Errol Flynn and Motorola Radio: Handie-Talkie Land with Paratroopers in "Objective Burma."' *LIFE*.

Noblis-Parks Industries. 1949. 'ARVON. New Super-Powered Portable Really Reaches Out!' *LIFE*, 2 May 30.

Philco. 1944a. 'The First Network! Another Milestone in the Progress of Television (Philadelphia-NY-Schenectady).' *LIFE*.

– 1944. 'From Radar Research to Radio for Your Home.' *LIFE*.

– 1947. 'Amazing New Philco Auto Radios.' *LIFE*, 17 Nov.

Pullman. 1947. 'Let's Go! And Here's How to Get There! Go Pullman.' *LIFE*, 28 April, 25.

RCA. 1947a. 'RCA: Radio Corporation of America RCA Laboratories – Your Magic Carpet to New Wonders of Radio and Television.' *LIFE*, 27 Oct., 12.

– 1947b. 'RCA: Radio Corporation of America RCA Radar – Enables Ships to See through Fog, Darkness, Storms.' *LIFE*, 14 July, 83.

– 1947c. Teleran – Radio Eyes for Blind Flying (Teleran Pictures – Air Traffic Control by Radar plus Television!' *LIFE*.

– 1947d. 'FM Radio – another World in Listening Pleasure!' *LIFE*, 17 March, 12.

– 1947e. 'New FM – Noiseless as the Inside of a Vacuum Tube!' *LIFE*, 20 Oct. 26.

Stewart-Warner. 1947. 'Home Means so Much More When the Whole Family Enjoys This Thrilling Entertainment! Wonder Window Television.' *LIFE*.

Trans Western Airline (TWA). 1947. 'Airline Radar is Here!' *LIFE*, 2 June.

Trav-ler. 1947. 'Travel with Trav-ler: Listen … and Be Carried Away!' *LIFE*, 3 June.

Western Electric. 1947. 'Radar Puts the Finger on Our enemies! Western Electric. In Peace … Source of Supply for Bell System; in War … Arsenal of Communications Equipments.' *LIFE*.

Western Union Telegram. 1947. 'New Radio Wings Can Speed 2000 Telegrams at Once.' *LIFE*, 20 Jan., 23.

4 Atmospheres of Communication

JENNIFER GABRYS

'Active Air'

From the Wireless Fields of Poldhu in Cornwall, to Signal Hill in St John's, Newfoundland, the first transatlantic wireless telegraph transmission took place from a temporary fan aerial to a distant kite. Guglielmo Marconi had boarded a ship to Canada in the middle of the winter to receive the invisible transmission of three telegraphic dots that formed one letter: S. After several failed attempts, Marconi declared the transmission successful on 12 December 1901, even as detractors suggested he was simply diving the static. Crackling and popping through the atmosphere, and travelling thousands of miles from their source, the dots were scarcely audible. Yet this faint detection proved that wireless waves could both travel without cables and also issue beyond the curvature of the Earth.[1] 'Dot ... dot ... dot' then arrived as a curious precipitation, signalling not just the extended distance over which messages could travel, but also new formations in the atmosphere of communications.

Marconi's experimental broadcasts (together with numerous other developments in wireless taking place at the time)[2] contributed to the rapid ascendancy of wireless transmissions. Yet the wireless waves that ping across the electromagnetic spectrum today require distinct apparatuses for their transmission and reception: telegraphs, radios, televisions, telexes, radars, satellites, mobile phones, and wireless computer networks. While these devices are often considered the 'medium' of wireless (with the radio set referred to as the 'wireless' for some time), in fact wireless is the mode of communication that, as the definition goes, 'does not require a medium of transport.' Of course, this refers literally to transmission without intermediary cables or wires. The wireless receiver or transmitter stands in for the 'medium' of wireless, so that the space through which signals travel is apparently medium-less. But wireless signals draw our attention to this space in between, the atmosphere through which wireless waves travel, the intervening medium of the air.

The Wireless Fields where Marconi's signal first sparked are now a barely legible ruin, comprised of slumping foundations and a sea-worn monument. Yet what drifts more suggestively through this space are the wide sky and ocean, those spaces of traversal and resonance that were drawn together through the first wireless transmissions. The wireless ruins draw attention to this horizon, and the atmosphere through which wireless signals made their migration. At one time, the space of communication was imagined as an etheric expanse, a medium of its own that exerted a pull upon whatever travelled through its elastic force fields. Even when the ether was scientifically rejected, there remained a language and imagination for describing this hazy space where messages and energy accumulate and transfer. Today, this language and imagination continue to have relevance. Information, as architect Toyo Ito notes, is 'active air' (Dunne 1999, 26). This active air constitutes the medium and spatiality of communication – a spatiality that is atmospheric and dynamic. In many respects, communication – wireless or otherwise – exceeds the devices, interfaces, and wires through which we typically conceive of the *medium* of communication. Indeed, we find there is another medium, an atmospheric medium, through which we can divine more than dots. This chapter charts how that first wireless exchange of 'dot ... dot ... dot' relocates from the ocean to the city, and multiplies towards concentrations of wireless exchanges that give rise to expanded ecologies of transmission. This chapter then explores how an atmospheric mode of communication – like the ether, resonant and electric – delineates a much different type of urban space that gives rise to emanation, presence, and surround.

City of Sparks

From the time of telegraphy and radio, wireless signals have permeated the city. Exchanges among people and increasingly among machines take place through wireless 'clouds' of communication suspended over the city. While the language of networks may prevail in discussions of urban communication, increasingly more fluid metaphors, from clouds to liquid topologies, are emerging to describe the dynamic character of communication and mobility in the city.[3] As this chapter suggests, mobile and wireless communication in the city is atmospheric. In this sense, the wireless city is best understood through the drift and pull of its electromagnetic spectrum. From radio to sensor, distinct frequencies establish invisible circuits – not arcades and thoroughfares, but atmospheres of communication – that draw the city together as a space of multiple correspondences (Sheller 2004, 47). In the city of clouds and sparks, furthermore, we encounter fields of energy, or what Vilém Flusser calls 'flections.' 'When we are talking about a "new urbanism,"' Flusser writes, 'it is more useful to construct the image of the city as a field of flections' (323). The city contains zones of energy, registers of communication, mobility and magnetic attraction. These flections describe the energy and 'field of relations' through which the city 'gains contours.' Flusser further explains this city of intensities and correspondences:

> The relations among human beings are spun of differing densities on different places on the net. The denser they are, the more concrete they are. These dense places develop into wave-troughs in the field that we must imagine as oscillating back and forth. At these dense points, the knots move closer to one another; they actualize in opposition to one another. In wave-troughs of this type, the inherent possibilities of relationships among humans become more present. The wave-troughs exert an attraction on the surrounding field (including the gravitational field); ever more intersubjective relationships are drawn into them. Every wave is a flash point for the actualization of intersubjective virtualities. Such wave-troughs are called cities. (325–6)

The trough, an in-between space, is the magnetic space of relation – it not only exists between, it attracts.[4] These troughs, furthermore, might be described as spaces of communication, as atmospheres of wireless exchange.

As zones of energy, the city is then multiply located, surfacing through intensities of exchange. Overlaid on the hard grid of pavements and architectural edges, an urban weather collects, a weather of messages and connections. 'A striking aspect of this image of the city,' Flusser writes, 'is its immateriality.' Within these flections, 'there are neither houses nor squares nor temples that are recognizable, rather only a network of wires, a confusion of cables' (326). By allowing the usual hard and fixed image of the city to fade into the background, we can begin to take note of the ways in which seemingly more immaterial exchanges, as transferred through wireless devices or electronic media, alter the ratio and intensity of space and time in the city. These wireless frequencies mobilize more than just media and technologies – they mobilize orders of energy in the city.

Energy is a way of understanding the intensities of space and time. It is just such an intensive reading of electronic, or 'new,' media that Marshall McLuhan called for. We should not inquire into the workings of media and communication as discrete and linear operations, McLuhan suggested, but rather as intensive and environmental phenomena, or experiences of depth (McLuhan 1994c). This depth is atmospheric. An atmosphere is composed of intensive gradations. It drifts and fluctuates between clarity and noise. It becomes saturated and weighed down with pressure, a fog of messages. And it breaks, shifts with electric, lightning-like pulses. Wireless signals collect and transmit intensively, across electromagnetic frequencies. These frequencies draw together registers of space and time. The charged transmission of electric messages then assembles orders of space and time intensively, rather than extending with blank and infinite regularity. When attempting to locate ourselves within these electric, intensive, and even topological orders of the city, we further find that we must redraw our urban maps and courses of connection. In this respect, Flusser suggests that 'the new city is not geographically locatable,' but rather, 'it is everywhere where humans open up to one another' (327).

Since their inception, wireless technologies have stimulated speculation about the new topologies that emerge through previously unimaginable connections. Indeed, correspondence via the electromagnetic spectrum was bound to draw us into radically altered conceptions of space and time. Professor W.E. Ayrton, after reading Marconi's discussion of wireless technologies published in *Electrical Review* on 15 and 22 June 1901, made a statement before the Society of Arts in London about how we might locate ourselves – electromagnetically. Ayrton envisioned a time within the not too distant future, 'when if a person wanted to call a friend he knew not where, he would call in a loud, electromagnetic voice, heard by him who had the electromagnetic ear, silent to him who had it not. "Where are you?" he would say. A small reply would come, "I am at the bottom of a coal mine, or crossing the Andes, or in the middle of the Pacific." Or, perhaps, in spite of all the calling, no reply would come, and the person would then know that his friend was dead.' Such correspondence between electromagnetic organs across unfathomable distances seemed capable of spanning almost as far as the grave. Those endowed with these highly tuned organs could exchange messages that would be audible to no one else. So unreal did these conjectures seem at the time that Ayrton could only say that this was 'almost like dreamland and ghostland, not the ghostland of the heated imagination cultivated by the Psychical Society, but a real communication from a distance based on true physical laws' (1901, 820). When locating ourselves electromagnetically, we seem to inhabit some ghostly geography. But the ghosts, in this case, are real. They are the flickerings of an elusive, atmospheric spectrum. This spectrum, however ghostly, constitutes a space of extended inhabitation.

Spectral Ecologies

As we can see with these considerations of the spectral qualities of wireless transmissions, an atmospheric view of communication is not without precedent. Ideas about an ether of electrical or magnetic force were prevalent in the nineteenth century. The ether was originally conceived as a medium through which light or gravitational forces travelled. This 'material and vibratory medium' predated the discovery of electromagnetism, and it was understood to be the stabilizing and guiding invisible substance through which forces moved. The ether, as a stabilizing medium, only gradually fell out of favour after the delineation of the electromagnetic spectrum. Yet even with this dismissal, the ether remained a potent metaphoric device. Infused with poetic and energetic qualities, the ether was simultaneously a medium and an environment. It was an invisible yet all-encompassing atmosphere, constituting an 'undulating spatial foundation upon which the mobile contents of radiant energies were propped' (Clarke and Henderson 2002, 21). In this sense, it was even conceived of as jelly – as though all of space were 'filled with jelly' – and as an elastic medium through which energy and 'lines of force' travelled (21). This jelly, atmospheric broth, or elastic medium resonates with what Jeffrey Sconce discusses as the 'etheric "ocean"' of the nineteenth century. He writes, 'The advent of wireless at the turn of the century heralded a radically different vision of electronic presence, one that presented an entirely new metaphor of liquidity in telecommunications by replacing the concept of the individuated "stream" with that of the vast etheric "ocean"' (14).[5] Even as the ether came to be discredited, wireless technologies then gave renewed attention to an oceanic or atmospheric view of communications.

Indeed, while one version of the ether fell into disfavour, multiple other versions grew up in its place. Joe Milutis (2006) notes in his study on the ether that 'Hertz and Helmholtz found the ghost of the visible-light spectrum and translated the luminiferous ether of the nineteenth century into the electromagnetic spectrum of the twentieth' (38). The ether quickly transformed first into the electromagnetic spectrum and then into a series of 'standardized wavelengths' that were cordoned into designated uses by the U.S. Radio Act of 1911 (78). With this designation, the ether seemed to move closer to earthly territory. Indeed, the current delineation of the radio portion of the electromagnetic spectrum reads as an elaborate property map, drawn as it is in bands of varying width. Radio or wireless frequency allocations appear more as an urban grid, than as fuzzy zones of sky that correspond to a jumble of transmitters and receivers. Multi-modal and multi-mobile, the chart delineates a confluence of spaces, from sea to air and land; and a convergence of uses, from radio to mobile devices and weather satellites. Yet this 'spectral' space is perhaps best characterized, not as regions with sharp boundaries, but rather as overlapping, simultaneous, and even interfering electric atmospheres. The spectrum, with its electromagnetic fields of the sort that nineteeth-century physicists imagined, contains 'neighbourhoods' of electricity and populations of intensity (Luckhurst 2002, 75-88). These spectral neighbourhoods are less topographical and more topological. In this respect, Milutis suggests, 'one might want to call the ether (as distinct from the ownable plots of the electromagnetic spectrum or the unattainable vibrations known to the third eye) a fabric of signs that is both material and phantasmic, an electronic rain that is continuously decoded and received in common or poetic ways' (Milutis 2006, 85).

What is it that this material and phantasmic ether still allows, even when science declares its inaccuracy? What do these spectral ecologies enable that might otherwise be lost in the hard logic of frequency allocations? As Michel Serres (1982) suggests, the fuzzy space of the spectrum may allow us to open our eyes to the expansive, even atmospheric, space between previously conceived sharp boundaries. He writes:

The Devil or the Good Lord? Exclusion, inclusion? Thesis or antithesis? The answer is a spectrum, a band, a continuum. We will no longer answer with a simple yes or no to such questions of sides. Inside or outside? Between yes and no, between zero and one, an infinite number of values appear, and thus an infinite number of answers. Mathematicians call this new rigor 'fuzzy': fuzzy subsets, fuzzy topology. They should be thanked: we have needed this fuzziness for centuries. While waiting for it, we seemed to be playing the piano with boxing gloves on, in our world of stiff logic with our broad concepts. Our methods can now be fine-tuned and in the process, increased in number. (57)

Neighbourhoods within the electromagnetic spectrum are located not necessarily side by side, but through connection, emanation, and intensity. As McLuhan points out, electricity is, importantly, not about containment, but rather about relation and position between bodies. 'Again, as more is known about electrical "discharges" and energy, there is less and less tendency to speak of electricity as a thing that "flows" like water through a wire, or is "contained" in a battery. Rather, the tendency is to speak of electricity as painters speak of space; namely, that it is a variable condition that involves the special positions of two or more bodies' (McLuhan 1994a, 347). This variable and electric condition is atmospheric. It provides an expanded frame of reference for conceiving of communication – wireless and otherwise – in the city.

Atmospheric Medium, Wireless Milieu

The space that wireless communications seems to cancel from view – the ether, air, or atmosphere – then emerges here as a necessary area of investigation. This hazy, electric, and intermediary space is, by many accounts, the medium that enables us to communicate in the first place. As Serres (1982) writes, the atmosphere is at once an intervening and enabling medium. It is far from the airless world of black and white categories and simple systems, which is an 'imaginary world' only possible 'on the moon,' Serres argues, 'without any atmosphere' to provide the basis for differentiation, where 'no one can see a thing.' In this sense,

> the atmosphere, the air, the milieu (the medium), makes light diffuse; it outlines obstacles, lights the other side of walls, single-point light sources producing scallops and patterns. In order to have only light, one would have to live at the single-point light source, or the medium would have to be removed, creating a vacuum. As soon as the medium intervenes, the ray of light wanders about the world. We see only because we see badly. It works only because it works badly. Every system is a set of messages; in order to hear the message alone, one would have to be identical to the sender ... As soon as we are two, there is a medium between us, the light

ray is lost in the air, the message is lost in the interceptions, there is only a space of transformation. (69–70)

When Marconi discerned the 'dot ... dot ... dot' of the first transatlantic wireless transmission, he had to listen through the atmospheric static for a legible sign. Yet this same static that seemed to impede a clear reception was the very medium that allowed the transmission to take place. In the attempt to detect signals out of atmospheric static, Marconi further placed emphasis on the communicative equipment that continues to occupy the centre of attention today. Here were fan aerial and receiver, telegraphic signal and decoded message. The surrounding atmospheric medium may have fallen from view, but it has continued to lurk in the background as an inevitable aspect of communication. From media ecology to mediology, from mediasphere to media environment, various notions of a media surround have been employed to convey the idea that the medium does not begin and end with the screen, cable, box, or receiver. In this respect, Régis Debray (1996), who works through the concept of the 'mediasphere,' suggests the sphere of media cannot even be limited to something external, but is, again, something more topological. As Debray writes, 'Mediological man does not cohabitate with his technological surroundings, he is inhabited by his habitat; constructed by the niche he has constructed' (111). As this chapter suggests, the medium of wireless technology can be approached as just such a habitat, as a milieu that is as atmospheric as fixed.

The medium, as writers from Friedrich Kittler to Régis Debray have suggested, is a field of relations. Rosalind Krauss points to 'the medium's aggregate condition' as evidence of the difficulty of drawing a boundary around any medium. This aggregate or 'compound idea of the "apparatus" refers to all of the medium's supports: in the instance of film, from celluloid to projector, light, screen, and beyond (Krauss 1999).[6] But the apparatus, or medium, is even more than its raw material. It also includes economic, political, and social contexts; spatial and temporal registers; cultural practices and modes of circulation. These elements can never be scrubbed away from the planar dimensions of the interface, or from the seemingly innocuous glow of the latest gadget. While the 'medium' may acquire its distinctness by its instantiation and use – a radio, for instance, conceived of as a certain frequency, broadcasts, and a transmitter encased in a black plastic box with antenna and dial – it also inhabits a larger landscape that spans from the history of wireless to Clear Channel. In this sense, perhaps the distinctness of the radio as radio is not necessarily undone, but the idea that radio is only 'the radio' is.

The question that emerges is whether we should then locate the medium in the physical artefact, the screen, the message, the wires, the network, or all of the above. What the Wireless Fields indicate is that we may even reconsider the 'medium' as an assemblage or, in other words, as an atmospheric media landscape. Every formation of 'the medium' gives rise to a shifting media landscape, and delineates a dynamic space in and through which we forge our understanding of communication and its apparatuses. How these atmospheric landscapes are traversed, inhabited, and extended becomes the very basis for understanding the medium. In this respect, Debray (1996) writes, 'The error of futurologists and disappointment of futurists commonly arise from overestimating the *medium's* effect by underestimating the *milieu's* weighty plots' (16-17). In this estimation, the medium is at once a process of mediation and a middle space, an environment. When answering the question 'what is a mediasphere?' Debray suggests, 'the chronological unifier can be called the *mediasphere, or middle ground, setting or environment* [milieu] *of the transmission and carrying* [transport] *of messages and people*' (26). The medium, as milieu, importantly involves the exchange and mobility of signals. In this respect, Debray writes, 'A mediasphere's space is not objective but trajective. It would therefore be necessary to hazard the term "mediospace," the relation of a given surface area to a duration. The "ball of earth" as a mediospace of the graphosphere is not the same as that of the videosphere. The one has a circumference of three years (Magellan) and the other of twenty-four hours (Airbus)' (29). Clearly, the spheres, spaces, and milieu that Debray draws out have important connections to this inquiry into the atmosphere of communications in the wireless city. We arrive in this discussion not just at an awareness of the environment of communications, but also of its emergent and contingent properties. The *atmosphere* is quite literally the space through which wireless signals travel, but it is also the historic and poetic substance that has enabled the speculation towards communication without wires, as well as a social, political, and economic apparatus. An atmospheric construct further enables a sense of the relation between electromagnetic trajectories and new geographies, whether the 'ball of earth' or cities. Wireless technologies do not just connect spaces; they give rise to new and shifting spatial and temporal orders.

In addition to Debray's discussion of Magellan and Airbus above, we can then add such wireless spheres as Nikola Tesla's (1995) 'World

Wireless System.' Another early pioneer of wireless, Tesla was responsible for critical developments within wireless technology. Yet his proposition for a 'World Wireless System' ultimately fell short of commercial success or practical use. With his 'World Wireless System,' for instance, Tesla proposed nothing less than the wholesale excitation of the earth through the application of wireless energy. As he elaborates, this system

> makes possible not only the instantaneous and precise wireless transmission of any kind of signals, messages or characters, to all parts of the world, but also the inter-connection of the existing telegraph, telephone, and other signal stations without any change in their present equipment. By this means, for instance, a telephone subscriber here may call up and talk to any other subscriber on the Globe. An inexpensive receiver, not bigger than a watch, will enable him to listen anywhere, on land or sea, to a speech delivered or music played in some other place, however distant. These examples are cited merely to give an idea of the possibilities of this great scientific advance, which annihilates distance and makes that perfect natural conductor, the Earth, available for all the innumerable purposes which human ingenuity has found for a line-wire. (Tesla 1915, 87)

Tesla's system quite literally performs the mobilization of the earth in its entirety as a wireless electrical system. This excitable planet would be in comprehensive and instant communication, enshrouded in an atmosphere of signals (Tesla 1999). With this mobilization, matter becomes electrical, and the planet acquires a new climate of wireless energy. Following upon Michel Serres and Bruno Latour, Steven Connor remarks on the contemporary movement towards such conceptions of matter that are more volatile, 'gaseous,' and informational. 'If history is marked by the movements, not from element to element, but between different states of the same element,' Connor (2004) writes, 'then time (temps), as Serres often takes pleasure in pointing out, becomes indistinguishable from temperature – or weather (temps). Solidity is just another way of naming speed ...' (105-17). With Tesla's 'World Wireless System,' planetary time has condensed to a matter of seconds. With this condensation comes an increasingly atmospheric conception of matter. Wireless technologies facilitate the movement towards fluidity, speed, and instantaneity. By extension, the time or temps of wireless then has a distinctly atmospheric temps or weather. How can we begin to discuss this distinctive weather in the wireless city to encompass the gaseous, the mercurial, and the meteorological?

Data Clouds

It is no accident that urban wireless networks put in place today bear the name of 'cloud networks.' One of the primary wireless providers in London, which charges for its service, is known as The Cloud, or Cloud Network. The Cloud is currently extending its atmospheric coverage over London by developing a citywide system of transmitters in lamp-posts. While these clouds may come at a price, they move towards a more localized version of Tesla's proposal for an expansive wireless system. With the development of these clouds, the city acquires an atmosphere of communications, a weather of signals. McLuhan suggests that early developments in the telegraph actually enabled the further refinement of weather forecasts, and that in many ways this technology enabled a new attention to 'weather dynamics.' As McLuhan (1994b) writes, 'It is clear that telegraph, by providing a wide sweep of instant information, could reveal meteorological patterns of force quite beyond observation by pre-electric man' (257). In many ways, this statement reveals how the telegraph not only brought attention to the weather, but also created weather, and came to operate as weather. Instant, electric, and global, telegraphic and wireless signals offered improved means for monitoring shifting and dynamic climatic phenomena because they also were shifting and dynamic technologies of atmospheric proportions.

Clouds then become apt descriptors of the wireless city on many levels. Clouds appear to be at once material and immaterial, emerging through a simultaneous process of formation and dissolution. They are visible only to become invisible, spectres of transformation. Airborne and ephemeral, they also graze structures and deposit residue. Clouds layer onto other architectures, moving through and transforming these seemingly impermeable forms. All that appears immobile at another level mobilizes through the transfer of energy. Hubert Damisch (2002) aptly describes these qualities of clouds through a discussion of painting when he writes, 'On a conceptual level, a "cloud" is an unstable formation with no definite outline or color and yet that possesses the powers of a material in which any kind of figure may appear and then vanish' (31). These clouds often stand in contrast to the representational devices of fixed perspective, because 'the sky does not occupy a place, and cannot be measured; and as for clouds, nor can their outlines be fixed or their shapes analyzed in terms of surfaces. A cloud belongs to the class of "bodies without surfaces," as Leonardo da Vinci was to put it, bodies that have no precise form or extremities and whose limits interpenetrate with those of other clouds' (124). Like the remote reverberation of wireless signals, clouds are beyond perspective and elude the fixed horizon; they travel in a space that does not exclude but is neither settled into lines. In many respects, the drift of these clouds – wireless, electric, informational – requires that we reconsider the city through the air. Instead of fixing on the pavement, we can begin to consider the more atmospheric transmissions and dynamic relations of cities.

The Public Electric

Through these more atmospheric dynamics, moreover, we can gain insight into the shifting publics that emerge in the wireless city. As Mimi Sheller (2004) suggests, 'Publics, in this formulation, are special moments or spaces of social opening that allow actors to switch from one setting to another, and slip from one kind of temporal focus to another' (48). Such switching and mobility reveal yet another aspect of the atmosphere of wireless communications, where the weather of messages provides access to a collective sense. This collective sense emerges in many discussions of media, electricity, and the ether.[7] McLuhan makes frequent reference to a media sensorium, where the 'central nervous system' is 'outered' to become a technological field (McLuhan 1994b, 247). So, too, does Milutis (2006) refer to the 'electric sensorium' of the ether (78). These electric and mediatized sensoriums in many ways are attempts to draw together a space of collective sensation that is persistently elusive. The electric sensorium is just as nebulous as the ether, a charged space of electrical storms that must draw us together into some atmospheric exchange. This electric sense is the sixth sense; it is that electrical sensation that reportedly we once had but have since lost. While animals such as sharks have a distinct ability to detect and respond to electrical fields, we can only conjecture through the shadow of sensory memory what the effects and trajectories of wireless signals and electricity induce. Our mobile and wireless devices may allow a dim prosthetic access to this electric sense, but the public electric remains largely a project of the imagination. But this is not necessarily a bad thing. Perhaps it is exactly these moments of imaginative induction that give rise to considerations of where public space is located in the spectrum of the wireless city: not just as a delineated frequency, but as a necessary interpretation at the juncture of multiple and complex social and technological processes.

In the early development of wireless, another inventor working simultaneously to Marconi took a much different approach to this technology. J.C. Bose, an Indian physicist working in Calcutta, began orchestrating wireless effects in 1894, at the same time that Tesla was making proposals for wireless and radio communication. Using wireless transmissions of electromagnetic waves, Bose sent sparks through gunpowder and rang bells at a distance. Bose went on to meet Marconi, but he was deliberately not interested in developing wireless for commercial use.[8] In the work of Bose, the spectrum remained an open space, a commons for the electromagnetic public. The spectrum as commons is perhaps a much less popular notion today, even though it seems self-evident on many levels that nothing could be so public as the air. Writing on the public aspect of radio, Gillian Beer (1996) notes that 'radio produced a new idea of the public, one far more intermixed, promiscuous and democratic than the book could cater for' (150). The unimpeded storm of messages travelling over the airwaves assembles as a space of potential connectivity, a space 'we switch in and out of' (149). On the airwaves there exist potential publics that can shift, assemble, and disperse at any time. This spectral commons reconfigures the city to suggest that we no longer map the virtual or physical, but rather register the intensity of electric atmospheres of communication. Indeed, Sheller (2004) writes, 'publics are no longer usefully envisioned as the open spaces or free spaces in which diverse participants could gather,' but rather 'the capacity for publics to emerge remains a property of the structures of connectivity' (50).

The technological medium operates as a charged electrical environment that informs how urban spaces and publics emerge. New public spaces and actions emerge through the spectrum, whether on the 'amateur' band or at proliferating sites of transmission and reception. Such a conception resonates with another set of atmospheres – the 'Atmospheres of Democracy' explored by Bruno Latour and Peter Weibel in the recent 'Making Things Public' exhibit at the ZKM in Karlsruhe. In the catalogue for this exhibition, Latour explains that by investigating the atmospheric qualities of democratic assembly, the show attempts to understand the more fleeting or even 'phantom' qualities of publics. Publics and public space are not only mobile, they are also potentially transitory, formed through shifting assemblages of 'things' or issues of concern that are continually coming into being (Latour 2005). To make things public is an atmospheric concept and practice. As this chapter suggests, atmospheric modes of communication open up spaces for thinking through the energies and possibilities of these public assemblies. In the wireless milieu, intensive and imaginative ways of configuring publics – and cities – emerge as a distinct potential of this atmospheric mode of inquiry.

NOTES

1 Marconi reportedly stated that, at the time of the transmission, 'the chief question … was whether wireless waves would be stopped by the curvature of the Earth. All along, I had been convinced that this was not so. The first and final answer came at 12:30 when I heard … dot … dot … dot' (Briggs 2001).

2 As many historians and scholars of wireless technology have noted, Marconi did not necessarily 'invent' wireless technology, nor was he the only pioneer working in this field at the time. Many scientists across the world, including Nikola Tesla, J.C. Bose, and A.S. Popov, were involved with wireless developments. Marconi, however, is often considered the first person to achieve success with *commercial* applications of wireless technologies. See Hong, 2001 for further discussion of the rise of wireless technologies and contested lines of authorship.

3 For more on the contrast between network and fluid models of urban mobilities, see Sheller, 2004. As Sheller writes, 'Whereas networks connect smaller units into larger entities, and such entities in turn form their own networks which constitute still larger social organisations, a gel is something in which such levels are not distinct. If we understand socialities as always grounded in physical space and time, but in contexts of sheer messiness, we may need to think about social life in nonnetwork terms' (47).

4 Régis Debray similarly notes in his study on the mediasphere, 'It is in reality the intermediate spaces and time, the betweenness of two things or periods, the trough of the wave [les entre-deux], that are decisive; but our language works the opposite way: it spontaneously subordinates the signs of relation to those of being, and doing to being' (1996 [1994], 11).

5 See also Connor 2004 on similar conceptions of electricity as atmospheric 'effluvium' or networked circuitry.

6 See also Kittler 1991.

7 For more discussion on wireless technologies and the urban sensorium, see Gabrys 2007, 2010.

8 See Bose 1927.

PART TWO

Mobile Practices

5 Mobile Publics and Issues-Based Art and Design

ANNE GALLOWAY

Introduction

Despite increasing research, design, and artistic efforts towards sup-
porting a multitude of 'publics' in mobile, context-aware, and pervasive computing, little critical attention has been given to how 'public' is
being defined and used – or how it relates to historical understandings
and practices. Usually associated with broader cultural shifts in the
production and consumption of networked information, mobile technologies are often positioned as the 'next big thing' when it comes to
citizen participation and an apparently long-overdue revitalization of
democracy. The technological focus on location and connection also
directs most art and design (i.e., non-commercial or pre-competitive)
projects to matters of space and culture, which are also matters of politics and ethics. In these ways, 'public' is used to refer to any number of
overlapping and divergent institutions, infrastructures, spaces, practices, interests, and values.

But before I go any further, I would like to be clear about how I
understand relations between publicities and privacies. As British sociologist Mimi Sheller (2004, 39-40) explains, 'In normative democratic
theory, the separation of the public and the private ... is central to en-
visioning the bases of social cohesion, political participation, and democracy itself.' Or as noted American pragmatist John Dewey (1991
[1927], 15) more bluntly put it, 'The line between private and public is
to be drawn on the basis of the extent and scope of the consequences of
acts which are so important as to need control.'

According to the *Oxford English Dictionary*, both 'public' and 'private'
entered into the English language in the early 1400s, and have always

been understood as opposites. As such they stand as distinctly socio-political concepts, or more specifically as questions of 'juridical distinction' (Hannay 2005, 5). However, most contemporary discussions involve multiple understandings of what constitutes both public and private, and Mimi Sheller and John Urry (2003) table the following categories of use: public/private interest, public/private sphere, public/private life, public/private space, and publicity/privacy (110). In all these senses, public and private are positioned as opposing values, although individual categories should not be seen as mutually exclusive. There is also a related tension between individualism and collectivism within each of these categories, as 'private' citizens are meant to assemble into some sort of 'public' at least some of the time. This complexity and fluidity is usually described in terms of multiple publics and privates, or networked relations between different publicities and privacies. However, ultimately Sheller and Urry (2003) argue that

> the characteristic ways in which the public/private distinction has been drawn, and the overwhelming concern with the problem of 'erosion' of the public sphere or 'blurring of boundaries' between the public and the private, fail to capture the multiple mobile relationships between them, relationships that involve the complex and fluid hybridizing of public-and-private life. (108)

Their perspective led me to seek out a sense of 'public' that better embodies such a complex and fluid hybridity, and this chapter traces particular points through which I passed in my investigation into what a 'mobile public' might actually be.

Starting with the 'problem' of the public, I look to select historical and philosophical understandings of publics and politics. Building on the work of early American pragmatist Walter Lippmann, I focus on a public that is fragmented and contingent but still very much capable of judgment and action. In order to delve deeper into the kinds of situations or events in which these kinds of publics can come together, I find inspiration in the carnivals and feast crowds so eloquently described by Mikhail Bakhtin and Elias Canetti, as well as in Bruno Latour's 'parliament of things' or *dingpolitik*. I follow that discussion with an overview of recent research into the social and cultural aspects of mobile, context-aware, and pervasive computing, and I question the senses of 'public' and 'private' at play. More specifically, following Mimi Sheller, I ask what a non-network model of mobility might look like. The kind of fluid and messy picture that emerges ends up pivoting on acts of coupling and decoupling, or gelling and dissolving, multiple publics and privates around shared concerns or difficult issues. The chapter culminates in a discussion of what I call issues-based art and design, or those mobile and context-aware projects in which a 'public' is convened around a set of shared concerns or complex issues that cannot be adequately handled by more traditional means. More specifically, I look at mobile technologies being deployed in the interests of political and economic awareness and action, as well as environmental awareness and sustainability. Assessing the limitations and possibilities of these kinds of technological, artistic, and design interventions, I conclude by asking where the most productive potentials for mobile publics can be found, and what it will take to actually mobilize them.

The 'Problem' with the Public

In current English usage, the word 'public' most often refers to 'the people.' However, as philosopher Alastair Hannay (2005) explains at length, beginning in ancient Rome 'the public' was inexactly aligned with 'the people' as a political force. Put another way, 'the public' and 'the people' were not one-and-the-same, and the 'public good' referred more to that of the state than of its combined citizens – although ostensibly the state also existed as a means to enable and support individual citizenship. By the seventeenth and eighteenth centuries, Europe saw 'the public' more properly become 'a public' or, rather, many publics. Importantly, this new kind of audience 'consisted and consists of all those able or prepared to judge what they saw and can still see' (Hannay 2005, 27). This matter of judgment is also integral to German philosopher and sociologist Jürgen Habermas's (1989 [1962]) 'public sphere,' and is characterized by Enlightenment traditions of rational and informed dialogue and debate.

The idea of 'public' as both actor and audience has also long been connected to technological development, especially as related to media production and consumption. For example, in the mid-1800s Danish existentialist philosopher Søren Kierkegaard implicated mass media and communication technologies when he blamed 'the press' for turning 'the public' into a 'monstrous abstraction, an all-encompassing something that is nothing' (1978 [1846], 79). Kierkegaard's 'public' was indifferent, and people were left with an inability to act, which has serious consequences for matters of collective political and ethical agency. At the turn of the century, German sociologist Georg Simmel (2000 [1903], 3) conjured a similar kind of political impotence in his descriptions of the 'blasé attitude' and faster, increasingly technologized, urban life.

This sense of an abstract or 'phantom public' was most notably taken up by American philosopher Walter Lippmann in the 1920s, revitalized in the early 1990s in an edited volume called *The Phantom Public Sphere* (Robbins 1993), and more recently explored in Dutch philosopher Noortje Marres's (2005, 2006) discussions of issues and democracy. However, unlike Habermas, Lippmann did not believe there was a coherent 'public' that directed the course of events, and his research on the 'phantom public' articulated what he saw as the limits and complications of public life:

> The public is not, as I see it, a fixed body of individuals. It is merely those persons who are interested in an affair and can affect it only by supporting or opposing the actor ... It follows that the membership of the public is not fixed. It changes with the issue: the actors in one affair are the spectators of another, and men are continually passing back and forth between the field where they are executives and the field where they are members of a public. (Lippmann 1925, 77 and 110)

Lippmann believed that controversy would necessarily arise from these competing interests, and he saw the 'problem' with the public as one to do, in part, with articulating the means to evaluate or judge so many divergent positions. For example, in *The Phantom Public* Lippmann wrote that while some issues are plain and established rules provide clear solutions,

> in many fields of controversy the rule is not plain, or its validity is challenged; each party calls the other aggressor, each claims to be acting for the highest ideals of mankind ... Yet it is in controversies of this kind, the hardest controversies to disentangle, that the public is called in to judge, where the facts are most obscure, where precedents are lacking, where novelty and confusion pervade everything, the public in all its unfitness is compelled to make its most important decisions. The hardest problems are problems which institutions cannot handle. They are the public's problems. (Lippmann 1925, 131)

While at first glance this does seem to be an intractable problem, Marres (2005) points out that Lippmann's position can also be seen to reinvigorate agency and urgency in public politics:

... it is thus the failure of existing social groupings and institutions to settle an issue that sparks public involvement in politics. It is the *absence* of a community or institution to deal with the issue that makes public involvement in politics a necessity. Because if the public doesn't adopt the issue, no one will. (212)

Furthermore, Marres (2005) argues that Lippmann's public is 'precisely *not* a social community' (214). Or, more specifically, the 'community' is not pre-existing; it is created by particular people implicated by particular issues. However, rather than falling prey to impotent abstraction, these publics rally force in their inconsistency and contingency:

> ... the agency of the public derives in part from the fact that this entity is *not* fully traceable. That is, the force of the public has to do with the impossibility of knowing its exact potential ... The fact that the public cannot be definitively traced back to a limited number of identifiable sources is thus crucial to the effectiveness of the public: this is what endows publics with a *dangerous* kind of agency. (Marres 2006, 80)

The phantom public, then, has power precisely because its *potential* is both unknown and, in many ways, infinite. Its judgment may be situational, but it will be an actionable judgment nonetheless, and it will continue to congeal and dissolve new publics and judgments as needed – or, at least, as possible.

Practising Publics

Elements of individual and collective uncertainty, potential, and transition also appear in the work of Russian philosopher and literary critic Mikhail Bakhtin. Of particular interest here is how Bakhtin positioned action at the level of the practical everyday, and how reasoned ethics were seen to emerge from particular places and situations. These kinds of bottom-up ethics and micro-politics elicit something far more unstable and unruly than Habermas's ideal public sphere, and perhaps are more in line with the complex situations and issues conjured by Lippmann.

As Gardiner (2004) points out, where Habermas saw 'sober and virtuous debate' using 'ideal speech' in a 'public sphere,' Bakhtin witnessed in the carnival a 'tumultuous intermingling of diverse social groups and widely divergent styles and idioms of language ... including the use of parodic and satirical language, grotesque humour, and symbolic degradations and inversions' (38). Public gatherings like the carnival privileged radical difference and multiple voices in many of the same ways as Lippmann's publics, but also made clear that these kinds of coming-together can be transformatively transgressive:

> ... all were considered equal during carnival. Here, in the town square, a special form of free and familiar contact reigned among people who were usually divided by the barriers of caste, property, profession, and age. (Bakhtin 1984 [1965], 10)

'Feast crowds' were also described by Bulgarian-born Nobel laureate Elias Canetti as ones in which 'everyone near can partake' (1998 [1960], 62). Canetti's feasts, like Bakhtin's carnivals, are situations of difference, excess, and potential where 'there is no common identical goal which people have to try and attain together. The feast *is* the goal and they are there' (Canetti 1998 [1960], 62). Working within this metaphor, we can see feasting publics to be those gathered around shared objects and concerns – or, more specifically and in similar ways to Lippmann's publics, their 'goal' is to come (be) together. In other words, Canetti's feast crowds are dense with all sorts of different objects, rituals, and people that have similar disruptive potentials to Bakhtin's carnival and Lippmann's phantom public.

In *We Have Never Been Modern*, Latour (1993) posits a 'parliament of things' or gathering together of all parties (human and non-human) implicated in any given issue. Outlined more specifically in his call for a new politics of things, or *dingpolitik*, Latour (2005) notes that 'objects – taken as so many issues – bind all of us in ways that map out a public space profoundly different from what is usually recognized under the label of "the political"' (15). Latour's sense of the political also builds on the work of Lippmann, and he operationalizes this by asking us to stop expecting the 'body politic' to give us what it cannot – that is, adequate representation – and instead allow ourselves to be moved by the 'phantom public,' where 'politics will pass through [us] as a rather mysterious flow, just like a phantom' (Latour 2005, 38). While I do not agree with Latour that this process need be so mysterious, the idea that we enact politics, or that politics are enacted through us, can be a powerful (if somewhat ephemeral) metaphor to live by. Being moved by shared objects and concerns, and as Canetti claimed, moved 'not in one direction only,' is exactly the kind of situated publics Lippmann claimed we would need to take on the most difficult questions of their – and I would add our – times. Similarly, and also following Lippmann, Warner (2002) explains that a public is self-organized, a relation among strangers, constituted through attention and created through its own discourse. Far from *the public*, unified and total, Warner argues that multiple publics and counter-publics emerge and dissolve around, as well as struggle over, particular issues.

Mobile Technologies, Mobile Publics

If we continue to understand 'public' to comprise situational assemblages of people, places, objects, and ideas, then there is probably no area of technological research and development that better explores and exemplifies these complex relations than do recent activities in mobile and context-aware computing. Given the imperatives to locate and connect (see, for example, Galloway and Ward 2006, Green et al. 2005) that are embedded in otherwise diverse technologies, it should come as no surprise that today's wireless and wearable devices and applications offer unique glimpses into crucial sets of values and expectations surrounding 'the fate of the public.' However, as long as *mobile* publics are only defined in terms of *networked* publics, we risk oversimplifying complex relations in ways that may actually prevent certain kinds of political action.

Increasing calls within social and cultural studies for a spatial or 'mobilities' turn (see, for example, Bauman 2000, Cresswell 2006, Urry 2000) have been echoed in books like Hoete's *ROAM: Reader on the Aesthetics of Mobility* (2004) and Turner and Davenport's *Spaces, Spatiality and Technology* (2005). More to the topic at hand, social and cultural interests in mobile technologies have so far concentrated on local and global mobile phone usage, wireless infrastructure and pervasive computing, with exemplary research in anthropology (see Horst and Miller 2006, Ito et al. 2005), cultural studies (see Galloway 2004b, Goggin 2006, Mackenzie 2005), sociology (see Castells et al. 2006, Glotz et al. 2005, Katz 2006, Katz and Aakhus 2002, Kopomaa 2000, Ling 2004, Sheller and Urry 2006), social geography (see Graham and Marvin 2001), architecture (see McCullough 2004, Mitchell 2004), and computer-supported cooperative work research (see Brown et al. 2001, Hamill and Lasen 2005, Ling and Pedersen 2005), as well as recent multi-disciplinary volumes (see Kavoori and Arceneaux 2006, Seijdel 2006), technology design books (see Greenfield 2006, Sterling 2005), and more popular accounts (see Agar 2005, Levinson 2004, Rheingold 2002).

However, by returning our focus to matters of 'mobile publics,' it quickly becomes clear that relatively few of these publications actually tackle the question head-on. A recurring theme in mobile technologies discourse is the intrusion of 'private life' into 'public space' through the use of mobile phones, although there is substantially less engagement with any sort of 'reverse' process involving 'public' intrusions into 'private' life, with the exception of more recent surveillance studies (see, for example, Lyon 2003, 2006). Extending this to the realms of mobile art and locative media practice, we can also see a general and quite prolific focus on technological interventions in 'public space' requiring 'public participation.' However, most discussion and activity surrounding mobile publics as political forces employ notions of social and political networking, such as in the Annenberg Center for Communication's *Networked Publics* (http://netpublics.annenberg.edu/) research project, and network analyses such as those by Castells (2000) and Larsen et al. (2006). Other wireless commons research (see Schmidt and Townsend 2003) and public projects, including the Canadian *Mobile Digital Commons Network* (http://www.mdcn.ca/) and Montreal's Île Sans Fil (http://ilesansfil.org/) community Wi-Fi project, also rely heavily on network metaphors and networked urban infrastructures.

But before we can return to the notion of 'public' laid out earlier – one in which different people, objects, and ideas converge around shared concerns without the necessity of consensus or resolution – we may need to distance ourselves a bit from what Sheller (2004, 40) has described as the 'mathematically precise' or "hard" imagery' of networks, and focus instead on 'more liquid or messy social structures' and 'softer visions of porous sociality.' Put otherwise, the network model or metaphor is not well equipped to deal with uncertainty, inconsistency, and instability – conditions outlined as integral to the sense of 'public' put forth by Lippmann, Bakhtin, Canetti, Marres, Latour and Warner. However, discussions of mobility, liquidity, and flow (see Bauman 2000, Deleuze and Guattari 1983 [1972], Shields 1997, Urry 2000) offer alternative ways of understanding the kind of assemblages and assemblies at stake here. As Sheller (2004) concludes,

> It is the capacity for coupling and decoupling in various ways that enables social action and the emergence of persons … If 'persons' emerge as identities out of this social gel, it could likewise be argued that collective actors emerge in the same way – that is, as 'more or less rickety ensembles,' or sociotechnical assemblages, 'energised in some situation and style.' The

> mobilisation of publics, then, is not simply predicated on increasing the density or intensity of face-to-face ties (as in a network), but depends instead on the entire context of communication gelling, which enables momentary stabilisations of collective identities as publics ... Mobile publics can perhaps best be envisioned as capacitators for moving in and out of different social gels, including the capacity to take on an identity that is able to speak and to participate in specific contexts. (49-50)

This kind of temporary coming-together, gelling, or coupling around shared objects and concerns – or complex issues – is the kind of public and political agency put forth earlier in this chapter, and Sheller's advice to focus on those things that can act as capacity-builders leads us directly to what I call issues-based mobile art and design.

Issues-Based Mobile Art and Design

Given the limitations of using network models to understand mobile publics, and recalling Canetti's focus on the things that bring and temporarily keep different people together, as well as Latour's calls to assemble around shared concerns, I would like to take a closer look at specific art and design projects that are using mobile technologies to do just that. However, before continuing, I would also like to note that although a wide variety of spatial-annotation or place-based storytelling projects, such as *Cityspeak* (http://www.cspeak.net/), *MobileScout* (http://www.mobilescout.org/), *[murmur]* (http://murmurtoronto.ca/), *Regrets* (http://www.regrets.org.uk/), and *Yellow Arrow* (http://yellowarrow.net/), have the potential to raise, or bring forth, pertinent social and political issues through public speech, for the purpose of this discussion I will focus exclusively on mobile and context-aware technology projects that start with, or have already gathered, a particular political issue or set of concerns. I also want to acknowledge that while this selection is necessarily incomplete, it can nonetheless provide a valuable range of perspectives and practices upon which further research can build.

To begin, I would like to recognize recent developments in physical computing and take this notion of shared objects or things quite literally. American artist collective Finishing School's Public Interaction Objects (http://www.finishing-school.net/pio/index.html) project involves

> a series of low-tech participatory objects. They are physical interfaces engineered to create meaningful interaction with individuals in various

public contexts [and] ultimately promote curiosity and participation for those that come in contact with them. PIOs also challenge our all too prevalent responses of suspicion and alarm that many foreign objects create.

At its most fundamental level, then, the project takes issue with unknown or unfamiliar objects. This concern gets dispersed through multiple objects that serve as 'public' facilitators around further issues of foreignness, isolation, stagnancy, surveillance, disaster, and environmental sustainability. Combining elements of Bakhtin's carnival, Canetti's feast commodities, and Latour's 'object-oriented democracy' these 'public interaction objects' provide situated and material reasons for coming-together. The 'public' is not fixed, issues are complex, controversies are entangled, facts are obscured, and precedents are lacking – this is Lippmann's public, and Warner's publics and counter-publics.

A similar focus on material objects can be seen in the *MILK* project (http://milkproject.net/), by Esther Polak, Leva Auzina, and RIXC – Riga Center for New Media Culture:

> The MilkLine is one of the countless movements of the international food trade, in this case milk, produced by Latvian farmers, made into cheese by a local factory with the help of an Italian expert, transported to the Netherlands, stored in a charming Dutch cheese warehouse to ripen, sold at the Utrecht market and finally eaten by Dutch citizens …

While the objects (i.e., milk and cheese) and issues (i.e., international food trade) are quite clear, the matter of *human* publics and counter-publics continues to be unclear, as they end up being the most uncertain and ephemeral component of this particular mobile and hybrid public. Another example is Canadian artist Nancy Nisbet's *Exchange* project (http://www.finearts.ubc.ca/nisbet/new_work/exchange.htm), where, again, the objects (i.e., Nisbet's personal belongings, RFID tags) and the issues (NAFTA, border-crossing, national security, surveillance and identity) around and through which a 'public' emerges, are constantly shifting and so too is the human element.

However, most issues-based art and design is not so literally object-oriented, and less clear-cut issues emerge front-and-centre. For example, given the increasing availability and affordability of sensor technologies, as well as commonly hacked consumer electronics, several notable projects have explored different aspects of people's physical environments. Beatriz da Costa, Jamie Schulte, and Brooke Singer's project *AIR: Area's Immediate Reading* (http://pm-air.net/)

is designed to be a tool for individuals and groups to self identify pollution sources, [and] it also serves as a platform to discuss energy politics and their impact on environment, health and social groups in specific regions.

Proboscis, working with Natalie Jeremijenko, created *Robotic Feral Public Authoring* (http://socialtapestries.net/feralrobots/), which provides tools to

> engage people in evidence collecting in a fun and tactile way [and] enable people to feel they can learn about their environment and have the evidence to do something about it.

Tad Hirsch's *Tripwire* (http://web.media.mit.edu/~tad/htm/tripwire. html) project connects sensing activities to political action:

> Custom-built sensors hidden inside coconuts are hung from trees at several public locations to monitor noise produced by overflying aircraft. Detection of excessive aircraft noise triggers automated telephone calls to the airport's complaint line on behalf of the city's residents and wildlife. Documentation of noise incidents is archived for later analysis.

Although focusing on sound pollution instead of chemical pollution, the *Tripwire* project clearly provides an issue through which objects and people can gel, and actions occur. The primary difference between this project and the previous two is the composition of a 'public.' While an actual or present public is required in both *AIR* and *Robotic Feral Public Authoring*, the *Tripwire* project allows machines to do all the political work 'on behalf of the city's residents and wildlife.' Arguably, this sort of delegation is the kind of representative democracy that Lippmann and Latour considered so inadequate, but it is consistent with Warner's definitions of a public created through discursive action. In any case, Latour (2005) acknowledges that one of the ways we can effectively assemble is to deny handicaps – of body, language – and accept prostheses of all kinds, including machines.

This matter of more 'direct' political action involving pre-existing infrastructure is also explored in *Platforms* (http://theaphroditeproject .tv/saftey/), by The Aphrodite Project. The issue at hand is the safety of urban sex workers, and the project pivots on networked women's shoes:

> Each sandal will have an audible alarm system, which emits a piercing noise to scare off attackers. The shoes are also outfitted with a built in GPS

receiver and an emergency button that relays both the prostitute's location and a silent alarm signal to public emergency services. Where there are problematic relations with law enforcement ... the shoes will relay the signal to sex workers' rights groups.

Selecting an issue that has arguably been beyond the scope of traditional sex work problem-solving measures, *Platforms* leverages mobile and location-aware technologies to link together a wider variety of actors and create situational, dynamic, and active 'publics' around particular safety concerns.

Taking the idea of a 'mobile public' based on shared concerns most literally, the final project I would like to discuss is Lisa Lynch and Elena Razlogova's *Guantanamobile* (http://guantanamobile.org/). Not within the same category of mobile devices discussed so far, this project relies more precisely on automobility:

> *The Guantanamobile Project* is an attempt to both inform and collect public opinion. We believe it is vitally important to help the American public understand the legal, political and territorial issues surrounding the Guantanamo detentions ... *The Guantanamobile Project* has three primary components – a website which serves as an information and survey database and networking center; and a mobile 'Guantanamobile' that will circulate information, perform field research, and hold nightly projection events; and a documentary about the practice of wartime detentions at Guantanamo Bay.

By taking the show on the road, so to speak, the project sought to act as a kind of 'mobile, roaming public sphere,' but if we return our focus to the idea that mobile publics are in fact repeatedly 'coupled' and 'decoupled,' then its value is in bringing together a variety of different people and perspectives at each stop in the journey. Yet perhaps more interesting for the purposes of our discussion, the issue itself has not remained stable either. As a project weblog entry of 15 June 2006 states:

> Over the past few months, we've realized that the situation has changed so much – and in most ways, changed for the worse – that our documentary project needs some serious updating. We're in the process of trying to figure out who we might want to interview, and how to try to continue to tell a coherent story about Guantanamo in the wake of all that has happened over the past few months. (http://guantanamobile.org/project/archives/2006_06.html#002947)

This serves as a reminder that not only are mobile and context-aware technologies well suited to dealing with physically distributed multitudes and situated events, but also able to effectively *adapt* to changing circumstances and concerns, or change direction. In the case of *Guantanamobile*, new objects can be created and taken 'on tour' again and again.

However, these projects are not without particular limitations. The first problem we encounter if we are interested in social, cultural, and political change is that we currently lack sufficient documentation to make solid assessments that can clearly guide future efforts. With the notable exception of the *Robotic Feral Public Authoring* project (see Lane et al. 2006 as well as extensive documentation at http://socialtapestries .net/feralrobots/) and, to a lesser extent *The Guantanamobile Project* (Lynch and Razlogova 2006), no substantial documentation of the process or product is publicly available for any of these projects. Without knowing these details, the optimism in my above discussion needs to be qualified. In other words, we should not confuse desired or expected outcomes with actual outcomes, and more open and collaborative research is needed before we can confidently claim a *politically powerful* mobile public.

The second problem is related to this last point but is a bit trickier as it again brings into question the very notion of 'public.' Since most of these projects took place at isolated events, like art festivals, or other individual locations, we should rightly ask exactly what the constitution of these publics was. Put another way, what kinds of national, regional, class, gender, race, ethnic, religious, etc. differences were *actually* performed and represented? Socio-spatial inclusion and exclusion continue to be very powerful forces that shape our experiences of 'publics' and 'privates,' and no amount of intervention, technological or otherwise, will be able to build capacity for *completely* equal or inclusive assemblages. On the other hand, I believe that any project that can successfully engage publics and counter-publics will always create the potential for meaningful political action.

Summary and Conclusion

This chapter started by identifying the question of what might constitute a mobile public. In the first section, I presented various perspectives on publics and counter-publics – Lippmann's phantoms, Bakhtin's carnivals, Canetti's feasts, and Latour's parliaments of things – and I asked if these kinds of emergent and contingent publics appear in contemporary social and cultural research on mobile and pervasive computing. In the second section, I suggested that most current research is actually focused on networked publics, and I argued that a more fluid and messy mobile public, based on the ability to couple and decouple around different issues, better offers the kind of radical political potential I presented in the first section. In the third and final section, a selection of what I call issues-based art and design projects was used to further explore the kinds of social and material relations at play in the kinds of mobile publics I advocated in the previous sections. And while these projects were considered indicators of great potential, without more empirical and critical social and cultural research that is publicly accessible, it is my fear that artists, designers, and researchers will continue to be limited in their abilities to be part of any politically powerful 'mobile public.'

In closing, I would like to reiterate that the sense of public I called for in this chapter is always already fragmented and unstable in terms of its actors – be they people, objects, places, or ideas – and issues-related art, design, and research will be continuously challenged to reiterate shared concerns and foster new publics. While this does conjure a sort of never-ending story, there is hope in the admission that 'we are not done yet,' and I am confident in our combined abilities to not just 'make-do' but more profoundly make and do.

6 The Third Screen as Cultural Form in North America

JUDITH A. NICHOLSON

The so-called third screen is 'coming of age' this summer ...
– A.A. Cuneo, 'Marketers Get Serious about
the "Third Screen,"' *Advertising Age*, 11 July 2005, 6

In the two decades since wireless cellular service was launched in parts of Asia, North America, and Europe, cell-phone use has shifted from mouth, to thumb, to eye and spread around the world. In North America, where the mobile phone is called the cell phone, shifts from talking to texting, and lately to TV watching and other mobile 'screening' practices, approximate the popularization of first-generation (1G) analog cell phones in the 1980s, second-generation (2G and 2.5G) digital cell phones in the 1990s, and third-generation (3G) multimedia cell phones in the 2000s. The shift to what I am calling 'screening' is exemplified in North America by the growing use of the term 'third screen' for the cell phone. I am using screening to describe the gesture of holding up a cell phone to frame a shot or to view an image on the screen. What is the third screen in a 'society of the screen' (Manovich 2001, 94)? It mediates screening, or ways of seeing. But, seeing what?

Sprint Corporation coined 'third screen' in 2003 and defined it as the screen after TV and the computer for the U.S. launch of its TV-to-cell-phone service (Ives 2004, Nowlin 2005). Sprint's service was panned as little more than a 'slide show' because of its slow streaming rate (Kerschbaumer 2003, Smith 2004), but the idea of the third screen as mobile TV took hold as Sprint and other American and Canadian carriers upgraded their networks in order to offer synchronous and asynchronous broadcasts of TV to cell-phone viewers.[1] By 2005, the third screen's coming of age was declared.

This chapter posits that the third screen became a cultural form in North America between 2003 and 2006. We increasingly connected to one another through it, and we used it to shape and interpret our times. The third screen became the site for competing corporate, cultural, and regulatory definitions of broadcasting, an intensified culture of payment, rival formats, new mobile aesthetics, and overlapping screening practices. The latter were mobile TV viewing, micro-movie making, and citizen journalism. While the third screen was advanced in North America, concerns related to cell-phone use since the 1980s were played out anew, but in regards to screening rather than to texting and talking. These enduring concerns, which are examined in a growing canon of studies on cell-phone use, have to do with how the practice intersects with perceptions and experiences of space and time, individual and cultural identity, and risks to health and safety. These concerns all pertain to the potential and risks of mobile communication, for which cell-phone use has become a metonym. What ways of seeing mobile communication did the third screen extend and foreclose in North America between 2003 and 2006 as it was shaped as a cultural form?

Space, Time, and Cell-Phone Use

By naming the third screen as the screen after TV and the computer for the launch of its TV-to-cell-phone service, Sprint declared an evolution in TV viewing and interactivity. Sprint framed its announcement in a familiar myth of technological progress. Its service seemed more like technological regression, however, because of limited program choices for subscribers and a streaming rate of 2 frames per second, which was well below television's 25 to 30 frames (Kerschbaumer 2003). Sprint upgraded its network a year later, as did other wireless carriers across North America. Sprint's service continued to be panned even after it increased streaming rates to 15 frames (Smith 2004). Third screen became a synonym for mobile TV despite Sprint's flagging efforts.

Within two years of Sprint's launch, media observers would herald mobile TV as the 'new anywhere, anytime TV model' (Foroohar 2005). Sprint's myth of progress was extended through evoking the mantra of 'anytime anywhere' communication. The mantra has been used consistently for years, and somewhat indiscriminately, in media reports and in research to describe how cell-phone users in North America and elsewhere blur present and future time by engaging in practices, variously described as 'micro-coordination,' 'hyper-coordination,' 'approximeeting,' and the 'softening of time' (Ling and Haddon 2003, Ling and Yttri 2002, Plant 2002). These perceptions of how cell-phone use blurs present and future time have endured even as the practice of archiving special text messages that are 'gifted' has grown and made the cell phone as much a 'compass' for the future (Kopomaa 2000) as a 'container' for the past (Rivière 2005). The mantra has also been used to describe how cell-phone users colonize public spaces through creating personal 'phonespace' with their talking and texting (Townsend 2000), through nurturing intimate 'tele-circles' with family and friends while travelling locally, regionally, and globally (Kopomaa 2002, Sherry and Salvador 2002), and through shaping a 'third space' that is 'adjacent to yet outside of home and workplace' (Kopomaa 2000).

The mantra of anytime anywhere communication reaffirms the binary of public and private spaces and the sequence of past, present, and future time. A tacit correlation between the immobility of these spaces and times and the mobility of cell-phone use make cell-phone use meaningful as mobile communication. In what follows, I argue that in order to interpret how the third screen as mobile TV extended and foreclosed certain ways of seeing cell-phone use, space, and time, we must pay attention to mobilities and to immobilities, or what has changed and what has not.

The Third Screen as Mobile TV

The third screen as mobile TV gained widespread media coverage with the creation of the first 'mobisode' in 2004. Fox Mobile Entertainment coined 'mobisode,' a contraction of 'mobile' and 'episode,' and patented the neologism for the launch of 24 Conspiracy, a mobile TV spin-off of its hugely popular espionage TV show 24 (Markis 2005, Kennedy 2006, Foroohar 2005, Shaw 2005). American wireless carrier Verizon signed an agreement with Fox to make 24 Conspiracy available to its subscribers. While each hour-long episode of 24 covered one hour in the day of the character Jack Bauer, an operative in a fictional Los Angeles counterterrorism unit, each episode of 24 Conspiracy was a truncated one minute (Foroohar 2005, Kennedy 2006). Within months, 24 Conspiracy was reportedly scheduled for translation into six languages in order to make it available to wireless carriers in over thirty territories around the world (McNicoll 2005). Other U.S. cable channels such as Nickelodeon, Comedy Central, VH1, and MTV began developing mobisodes, and three-to-five-minute soap operas, or 'mobisoaps,' as well as other programming for mobile TV (Manly 2006, Noguchi 2006, Shaw 2005).

By 2005, American wireless carriers such as Verizon, Sprint, T-Mobile, and Cingular began offering synchronous and asynchronous broadcasts of programming from channels such as CNN, MSNBC, C-SPAN, Fox Sports, and ESPN (Balint 2005). Canadian wireless carriers such as Bell Mobility, Rogers Wireless, and Telus Mobility signed agreements with TV networks to gain access for their subscribers to channels such as CBC Newsworld, the Weather Network, Fox News, the Shopping Channel, MétéoMédia, and Le Réseau de l'Information (Andrews 2005, Cribb 2005).

Wireless carriers across North America were not the only players who began to offer content for mobile TV in 2005. Major League Baseball and the National Football League sought to profit from the third screen as mobile TV by drafting plans to provide game content directly to wireless carriers, which would in effect make them broadcasters (Savage and White 2005). It was also reported that independent media producers were creating made-for-mobile TV content that they would offer directly to wireless carriers, effectively bypassing TV networks (Shaw 2005, Kennedy 2006). In addition, Apple Computer released its video iPod, four years after releasing its first personal media player (PMP), the iPod, and launching its on-line music store, iTunes. Apple announced it was signing agreements with American networks to sell TV shows via iTunes to video iPod users. For $1.99 per episode, shows such as *Law and Order*, *Desperate Housewives*, and *Lost* could be downloaded commercial-free (Gnoffo 2005, Shields 2005). Americans could also view these programs via the Internet, but Canadians could not because of broadcast rights held in Canada by the CTV network (Taylor 2006). At about the same time that Apple made its announcements, the ABC network launched an on-line store where fans of its shows, such as the soap opera *All My Children* and the espionage series *Alias*, could download images from the shows to their cell phones (Musgrove 2005). For a one-time fee, a monthly fee, or, occasionally, for free, users of cell phones, PMPs, and PDAs (personal digital assistants) began to stream or download commercial content through wireless cellular networks, iTunes, or the Internet. Apple announced that in the first twenty-one days after the launch of the video iPod, one million downloads of music videos and TV shows were bought worldwide through iTunes (Taylor 2005). Fifteen million downloads were reported in the first six months after the launch (Kennedy 2006).[2] Although by 2005 cell-phone users were already accustomed to paying for a handset, service, and downloads of ringtones, games, and graphics, Apple was credited in media reports with marshalling a shift away from illegal file-sharing of music and other digitized content to a new 'culture of payment' for content downloaded legally ('Digital' 2006, Siklos 2006, Werts 2006). Despite the fact that different portable hand-held devices were used to access mobile TV, and that web broadcasting was growing with the popularity of video-sharing websites, the cell phone retained its label as the third screen. The cell phone remained the 'holy grail' of wireless hand-held devices for content producers because of the number of users in the United States and Canada (Kennedy 2006), which by 2005 was, respectively, approximately 207 million and 12 million (CTIA 2006, CWTA 2005-6).

With more choices for mobile TV, media observers predicted the end of 'appointment TV' in North America (Cribb 2005, Foroohar 2005) and declared the 'age of television in our pockets' (McNicoll 2005). Some observers suggested that even with cable TV and with recorders such as the VCR and Tivo, choices of what to watch, when to watch, and where to watch were not as flexible as with mobile TV (Foroohar 2005, Pogue 2006, Shecter 2006). Some predicted that as third-screen users time-shifted their TV viewing to personalized schedules, mobile TV would be watched during 'in-between moments,' for example, while waiting in line or while waiting for an appointment (Shaw 2005, Pogue 2006). They predicted also that users would space-shift their viewing to the 'third space' and watch mobile TV, for example, while commuting (ibid.). Any remaining interstices of 'dead time' would thus be filled with 'snacks of TV' (Grover 2005, Levy 2005, McNicoll 2005, Shaw 2005). The focus in such commentaries was on how the third screen as mobile TV would mediate more choice and mobility. Overlooked was the correlation between mobility and immobility that underpins perceptions of space, time, and cell-phone use. So, what changed, and what remained unchanged?

Screening History and Cell-Phone Use

'This was the year that television arrived on our cell phones – *finally*,' declared one media observer in 2005 (Taylor 2005). In such comments, the categories of past, present, and future time remained fixed through contrasting the flexibility of mobile TV with the scheduled broadcasts of appointment TV. Though these categories remained fixed, time was truncated within the third screen. The content of mobile TV was described as 'bite-size' programming (McNicoll 2005, Manly 2006), which reflected, deliberately or not, the digital composition of such programming and its condensed form. The latter had to do, in part, with the memory capacity of 3G cell phones and streaming capacity of 3G wireless networks. Some observers predicted that 3G streaming was 'a prelude to the construction of dedicated mobile-TV broadcast networks, which [would] transmit digital television signals on entirely different frequencies to those used for voice or data' ('Fuzzy Picture' 2006). In the meantime, users of the third screen streamed and downloaded live or already broadcast TV content for immediate or future viewing. Time was mixed up not only in the third screen but also in commentaries about it.

Sprint framed its announcement of mobile TV in a familiar myth of technological progress, but researchers of cell-phone use and media observers frequently remarked, and still do remark, that experiments with mobile TV began in 'gadget-loving Japan' ('Gadget' 2006) and in South Korea in 2001 (Goggin 2006, 181; Lacey 2005, McNicoll 2005, Shaw 2005). Some such comments perpetuated the 'techno-orientalism' (Ito 2005, 2) that runs rampant through much research and commentary on cell-phone use.[3] Such comments also gestured towards North America's past reputation as the site of early wireless innovations that led to the cell phone's unveiling in 1973 by Canadian engineer Martin Cooper, who was dubbed 'father' of the cell phone (Steinbock 2005, 44; Galambos and Abrahamson 2002, 32).

When Cooper made the first-ever cell-phone call from a New York City street on behalf of Motorola, he became the pivot in a genealogy of innovative men that stretched out behind him, back to the pre-cellular era of wireless innovations, mostly in the United States, early in the century (see Steinbock 2003, 2005), and also stretched out after him to the launch of the wireless cellular era in the mid-1980s (see Merriden 2003, Curwen 2002, Galambos and Abrahamson 2002, Murray 2001, Corr 2000, Surtees 1994).

In the 1990s, North America's reputation as the place where wireless innovations emerged was overshadowed by innovations in Japanese *keitai* design and services and concomitant attention to *keitai* use from researchers worldwide (Ito, Okabe, and Matsuda 2005, 2). While the third screen was advanced as mobile TV in North America, the past peeked through in comments about it. Exporting of the mobisode *24 Conspiracy* to other regions around the world augured North America's resurgence as a wireless innovator, ironically, at about the same time that the Abu Ghraib scandal broke and it was revealed that American soldiers stationed at the prison used 3G cell phones to make and circulate photos and videos of torture that harkened back to 'lynch carnivals' of a century past in the U.S. South. I will say more about the prison scandal in the last third of this chapter. As mobile TV, the third screen mediated past, present, and future viewing for its users. As a cultural form, the third screen mediated history in North America as well as the current conjuncture of war on terror.

In terms of space, the third screen as mobile TV drew attention to the space of the cell phone itself. As the third screen, the cell phone was seemingly remade as a visual mass medium. It was defined as mediating personalized viewing through being contrasted with standard TV. The latter was defined as domestic and collective viewing, though these characterizations have been challenged and, it could be said, even refuted (see McCarthy 2001, Morley 2000). The binary of public and private spaces remained fixed in interpretations of the third screen as mobile TV. The contrast established between mobile TV and standard TV echoed how early cell-phone use was distinguished from wired-phone use through the idiom of 'cutting the cord.' However, while cell-phone users created the voice and text message content for 1G and 2G cell phones, the content that flowed through 3G cell phones by 2005 also included commercial TV. Some observers suggested that the third screen was not a new visual mass medium but merely a remote control for cable TV (Gertzen 2005). They argued that standard TV was not dead but, rather, 'cementing its own appeal' through mobile TV (Werts 2006) and, by extension, reproducing already established relationships between viewers and media producers. In such comments, new formats such as the mobisode and mobisoap were interpreted as lure created by media producers seeking to make mobile TV as profitable as standard TV once was. In other words, the familiar quest for profit advanced the third screen. This was echoed in the comments of one media producer who said at a symposium on mobile broadcasting in Toronto in mid-2006 that 'the great thing about the mobile phone is that it's an easy way to extract money from people.' Players in mobile TV seemed mostly concerned with what business model would reap the highest profits and what content would succeed as the signature content, particularly with the desirable youth market that was frequently reported to be users aged fifteen to thirty-four, and males, in particular, because they are the heaviest users (Foroohar 2005, Lacey 2005).

Observers noted that while mobile TV was being advanced, standard TV was losing its audience, and also advertisers, mostly to the Internet

(Foroohar 2005, Hein and McClelland 2005). According to one media observer, 'With broadcasters short of money, and advertising revenues falling, the tiny screen has begun to look like a potential saviour for the entertainment industry' (Lacey 2005). In light of such comments, *24 Conspiracy* can be seen as both a spin-off of *24* and a trailer to attract viewers to *24*. Debate continued in 2006 among media observers regarding how advertisers and media producers would actually push ads through the third screen outside of programs that already included product placements ('Fuzzy Picture' 2006, Shecter 2006).

In 2006, mobile TV earned some legitimacy when the National Academy of Television Arts and Sciences created a new Emmy category to recognize programming for 'non-traditional viewing platforms' like the third screen (Manly 2006, 'Wireless TV' 2006). In the same year, the Canadian Radio-television and Telecommunications Association (CRTC) made the decision to exempt 'mobile television broadcasting' from federal regulation ('No CRTC' 2006). Wireless cellular providers and the Canadian Wireless Telecommunications Associated (CWTA) argued, and CRTC concurred, that mobile TV is 'unlikely to compete significantly with traditional television broadcasting services due to the limitations of the wireless technology, the battery life and small screen size of the handset, the poor image and audio quality and the type and range of programming choices offered ...' (CRTC 2006). Rather than simultaneous one-to-many broadcasting like appointment TV, mobile TV is one-to-one broadcasting across a wireless cellular network or individual downloading from the Internet. The CRTC equated mobile TV with web broadcasting, which was already exempt from regulation under the 1999 New Media Exemption Order.[4] The CRTC essentially regarded the third screen as being identical to the computer screen. The Alliance of Canadian Cinema, Television and Radio Artists (ACTRA) saw the third screen as a potential threat to Canadian culture. Fearing that more American snack TV would flow to Canadian users of the third screen, ACTRA called CRTC's ruling a 'disappointment' ('Regulator' 2006). The American broadcast regulator, the Federal Communications Commission (FCC), had already decided in 2004 not to make policies for mobile TV (White 2004). The chairman of the FCC declared that mobile TV should be left to develop further before a decision was made about regulation (ibid.).

Whether repurposed or new, mobile TV content became the 'long tail' (Anderson 2006) of standard TV programming.[5] The third screen as mobile TV became yet another screening platform in a 'multiplatform media universe' (Crane 2006) dominated by global media conglomerates such as Fox. Micro-movie making seemed to challenge the commercialization of the third screen as mobile TV, but it was a muted challenge.

Cell-Phone Use, Identity, and Culture

In mid-2006, as I sat in a theatre in Toronto and watched micro-movies submitted to *Mobifest*, contradictions of the third screen played out before me. *Mobifest* was billed as the first mobile film festival in Canada. A small audience had gathered to watch the best of over two hundred entries. The micro-movies were still available for viewing on the festival website as they had been in the weeks leading up to the screening, when viewers cast their vote for the best. The independently made movies were doubly framed and branded as they were screened. They played within the projected faceplate of a cell phone manufactured by the festival's sponsor, Palm. The framing created the impression of looking, not at a film screen or a TV screen, but at a computer screen with several windows open simultaneously.

During the awards ceremony that followed the screening, festival co-host Mark McKinney, who is best known for his work with the comedy troupe Kids in the Hall, joked that the cell phone provided to him by Palm was 'the second most fun technology I've come across that fits into my pants.' McKinney's phallic quip earned a big laugh from the mostly young audience. Co-host and fellow funny guy Seán Cullen wisecracked, 'This is the ideal platform for Canadian film because we have no money!' Several times during the evening, co-hosts and presenters remarked that the third screen awaits its Steven Spielberg, an entreaty which also appeared in media reports (Kennedy 2006).

The evening at *Mobifest* established that even as commercialization defined mobile TV and framed the exhibition of independent micro-movies, users of the third screen were not just viewers; they were also creators. Naming Steven Spielberg as a director to emulate was a nod to the desirable young male market. Invoking Spielberg also demonstrated a new regard for users as creators, recalled the early years of cell-phone use when mostly men were users, and highlighted the continued dominance of American icons and culture in Canada. Issues regarding identity and culture have been another major focus of studies on cell-phone use in North America and elsewhere.

Early cell phones and cellular service in North America were very expensive, and, as a result, the cell phone was regarded as a business technology, 'rich man's toy,' and yuppie accessory (Katz and Aspden

1999, 49; Katz 1997, 236; Rakow and Navarro 1993, 148). Studies of cell phone use and identity in North America and elsewhere initially focused on who were the 'early adopters' talking on cell phones (Leung and Wei 1999; Katz and Aspden 1999, 49; Katz 1997, 236). These studies were followed by others on gender, the most critical of which focused on how cell-phone use by women both reproduced and challenged existing gender inequities (Rakow and Navarro 1993). After gender, studies focused on youth and young adults, particularly in Europe and Asia. Researchers examined how these users, dubbed 'thumb tribes' (Rheingold 2002) and 'generation txt' (Rafael 2003), used cell phones, and especially text messaging, to shape identity and social networks independently from their family (Kato et al. 2005, Green 2003, Ling and Yttri 2002, Nafus and Tracey 2002).[6] Some researchers have also focused on interpreting how media representations influence perceptions of cell-phone users, particularly sexist and racist perceptions (Shade 2007, Heckman 2006, Whitney 2006).

These studies, though admittedly varied, were concerned with how different forms and representations of cell-phone use were taken up by various groups in ways that challenged or reinforced existing identity and culture. In what follows, I argue that in order to interpret how the third screen as movie screen extended and foreclosed certain ways of seeing cell-phone use, identity, and culture, we must pay attention to who used the third screen as a movie screen to shape identity and culture.

The Third Screen as Movie Screen

Micro-movies used the third screen as a lens and, thus, challenged Sprint's exclusion of video and film in its definition of the third screen as the screen after TV and the computer. In Canada, as media reports on mobile TV began to appear frequently in 2005, so too did reports on micro-movies, which were also called 'mini-movies,' 'mobile movies,' and 'phone flicks' (Lewis 2005, Parks 2005). These various terms highlighted the newness of using the third screen as a lens. By 2006, this screening practice would burgeon in conjunction with citizen journalism and the popularity of YouTube.

In Canada, the first reports on micro-movies focused on *Shorts in Motion: The Art of Seduction*, an initiative by Bravo!FACT and the National Film Board of Canada (NFB). Original micro-movies were commissioned from four Canadians: writer-actor Don McKellar, filmmaker Sudz Sutherland, actor and radio host Sook Yin Lee, and actor-comedian

Mark McKinney. Reports on the initiative focused mostly on the 'mobile aesthetics' displayed in the micro-movies.

Using a Palm smartphone (Whyte 2005), McKellar shot a two-part micro-movie, entitled *Phone Call from Imaginary Girlfriend: Ankara* and *Phone Call from Imaginary Girlfriend: Istanbul*, both three minutes long. McKellar was quoted as saying that 'every century has had its version of miniature art, and mobile movies could be this century's' (Evans 2006). After seeing his micro-movie, which could best be described as an electronic soft-porn postcard, it becomes clear why McKellar also said he thinks of micro-movies as 'a little devotional thing to stare at in private' (Evans 2006). Viewing it at *Mobifest* in the GUI (graphic user interface) framing of the corporate screen, I was reminded that the Internet's most profitable content is porn and that in Europe, where 3G cell-phone use is more advanced, 3G is sometimes jokingly said to represent access to 'girls, gambling, and games' (Harkin 2003, 38).

Lee's four-minute micro-movie, *Unlocked*, is a sad and funny tale about a woman's dilemma over a relationship and a locked bicycle. Lee said she 'chose to film a lot of the action close up so that it would register well on a tiny phone screen' (Dixon 2005). Sutherland's four-minute micro-movie, *Go Limp*, is a quirky story of a procrastinating office worker. Sutherland said of micro-movie making, 'What can be transmitted and understood had to be very simple, like an old silent movie' (Whyte 2005). McKinney's six-minute micro-movie, *I'm Sorry*, asks one question of passersby: 'Is there anyone you owe an apology to?' The responses from a multi-generational and multi-racial group of respondents, who are shown mostly in close-ups, are confessional, funny, and touching. The micro-movies were broadcast on Bravo! (Lewis 2005), screened at the 2005 Toronto International Film Festival (Dixon 2005, Lacey 2005, Whyte 2005), and made available for download from the NFB website and *Shorts in Motion* website. I viewed McKinney's and McKellar's micro-movies at *Mobifest*.

Festivals celebrating the 'cell-ver screen' (Lewis 2005) or 'cell phone cinema' ('Coming' 2004) had been held in North America and Europe since 2003 (Lewis 2005, McNicoll 2005, Parks 2005). *Mobifest* was the first in Canada. It was sponsored by Palm, the maker of the Treo Smartphone, a combined cell phone and PDA. Mobifest received over two hundred entries from across Canada and around the world ('First Annual' 2006). Festival guidelines had stipulated that entries should be no longer than sixty seconds, to keep files to a manageable size for downloading ('Festival Terms' 2006). The guidelines had also cautioned entrants not to depend on sound to tell their story but, instead, to use titles to convey a story line, close-ups instead of wide shots for easier viewing on a small screen, and funny stories instead of sad ones to garner viewers' votes ('Making' 2006). The micro-movies varied from hilarious to dogmatic.

Entries were chosen for competition by viewers who cast votes after watching the micro-movies on-line or downloading them to a cell phone, PDA, or smartphone. This mimicry of democracy as audience interactivity had been popularized in Europe and North America by reality TV shows, such as *Popstars, Big Brother, Pop Idol, American Idol, Canadian Idol*, and *Star Académie*, which invited viewers to vote for contestants by calling in or text messaging. At *Mobifest*, prizes were awarded in three categories. Final judging was based on 'mobile aesthetics, audience impact and degree of creativity' ('First Annual' 2006). Coincidentally, while independent micro-movies were screened at *Mobifest* and at other such festivals, Hollywood producers began making shorter trailers that resembled micro-movies (Savage and White 2005). The trailers could be viewed on the Internet or downloaded and viewed on the third screen.

As these examples show, advancement of the third screen as movie screen opened the way for further commercialization, and it also opened the way for users to become creators. Micro-movie making did little to unsettle existing perceptions of cell-phone use as the domain of men. However, micro-movies further shaped new mobile aesthetics for the third screen and extended the practice of making cell-phone use a site of struggle over identity, and particularly cultural identity.

Screening Culture and Cell-Phone Use

Micro-movie making and mobile TV shaped a mobile aesthetics that contributed to the third screen's development as a cultural form. The emphasis on the use of visuals and text over aurality underscored the rise of screening practices. These aesthetics also mirrored web broadcasting and made the third screen seem like a computer screen. The double framing of micro-movies at *Mobifest*, and the tiny luminous cell-phone screens in the darkened theatre, brought to mind Lev Manovich's definition of the modern GUI as 'the coexistence of a number of overlapping windows' (Manovich 2001, 97). Some media observers suggested that micro-movies were a new form of movie-making (Manly 2006). Others were dismissive. Micro-movies were described as 'animated jokes,' 'teasers,' and 'quick gags' (Evans 2006, Lacey 2005).

NFB's involvement in *Shorts in Motion* gave the third screen as movie screen a certain degree of legitimacy. NFB's reputation of representing Canada to Canadians, of facilitating myth and knowledge about the nation through film and video, made the third screen like a 'third skin.' This metaphor for technologies that mediate mobile communication, such as the cell phone and the car (Frascara 2003), recalls Raymond Williams's use of car travel as a metaphor for 'mobile privatisation.' Williams coined mobile privatization to describe a structure of feeling, ideas, and emotions that he argued shaped and was shaped in part by car travel and by TV broadcasting in post-war Britain. Car travel facilitated corporeal and geographic mobility as well as upward mobility through the purchase of a vehicle. Television facilitated upward mobility via consumption of a TV set and facilitated what we now call virtual mobility through broadcasting. Mobile privatization described seemingly contradictory ways of living separately yet together, individually yet collectively, privately yet publicly, immobile in cars and living rooms yet mobile via roads and broadcasting (Williams 1992 [1974], 20). These ways of living exist in contemporary North American society, but now the third screen moves with us. The third screen has become our third skin. It is used to facilitate virtual mobility and to display symbolic mobility and power. All of these elements, and particularly the display of power, were evident at *Mobifest*. The display affirmed that the third screen is the cultural form of our patriarchal and white-dominated society.

Attending *Mobifest* felt like venturing into a men's club. Many of the micro-movies submitted to *Mobifest* were created by men, the audience was mostly male, the two co-hosts were male, the three judges were male, and the three award winners were male. Researchers have posited in studies of cell-phone use and identity that the cell phone is not just a status symbol for some men; it is a phallic symbol (Lycett and Dunbar 2000, Plant 2002, Roos 1993, Townsend 2000). This association was tangible at *Mobifest*. It was evoked in McKinney's phallic quip, in references to Steven Spielberg, and in the gaze of McKellar's micro-movie. The association mirrored the early years of cell-phone use in North America and brought to mind the Abu Ghraib scandal, in which the cell phone became a phallic symbol of the power wielded by U.S. soldiers and their nation in Iraq.

We have become increasingly connected to one another through the third screen in recent years. It has been used to reflect and interpret our times. This was evident furthermore in the mid-2000s with citizen journalism, the third screening practice that was popularized concurrently with mobile TV and micro-movie making. Citizen journalism was credited with exposing torture at Abu Ghraib.

Risk and Cell-Phone Use

In April 2004, news reports revealed that American soldiers stationed at Abu Ghraib Prison just outside the Iraqi capital of Baghdad had used cell phones and digital cameras to make over 1,800 photos and short videos in the previous months (Rajiva 2005, 12; Razack 2005, 348). The images showed soldiers torturing and humiliating Iraqi male and female detainees and posing with the battered corpses of detainees who did not survive the torture. A soldier who was not involved in the torture leaked the first images from the prison (Danner 2004, 215). The leak was called citizen journalism (Goggin 2006, 147; Robinson and Robison 2006, 90).

Citizen journalism has become a news staple. The term now usually describes when people who are not trained as journalists use cell phones, cameras, and camcorders to record eyewitness reports of momentous events and then transfer their reports and images to the Internet and other media for broadcast or publication and, sometimes also to friends and acquaintances. Though the soldier who leaked the first images from Abu Ghraib did not take them, his act was named as citizen journalism because he became the link between torture that was documented, but unseen, and the eventual worldwide broadcast of the evidence.

Citizen journalism was named after the September 11 attacks in the United States (Robinson and Robison 2006, Allan 2002), but its roots are often located in the 1991 video recording of Los Angeles police assaulting black motorist Rodney King (Robinson and Robison 2006). Citizen journalism was popularized as a screening practice in the mid-2000s in conjunction with the growing ubiquity of wireless hand-held devices and, specifically, in wake of a devastating tsunami in Southeast Asia, terrorist bombings in Madrid and London in 2004 and 2005, and the hanging of Saddam Hussein in 2006. Images recorded by witnesses of these events were broadcast worldwide. Through citizen journalism, cell-phone use intersected again with risk and safety, which has been another area of focus for studies on cell-phone use since the 1990s.

In the 1980s, cell-phone use was often promoted as a lifeline in emergency situations (Arceneaux 2005). Cell-phone use itself became the focus of concern about risk and safety in North America in 1993 after a

Florida man sued a phone manufacturer on the basis of his belief that radiation from a cell-phone caused his wife's brain tumour and her untimely death. Concerns faded in light of the lawsuit's dismissal (Burgess 2004), but the unproven link between cell-phone use and cancer was mentioned subsequently in many studies, however briefly, as a noteworthy moment in the early history of cell-phone use. Through hindsight, researchers declared that the cancer scare represented a vague 'discomfort over a new technology that crystallized into a specific fear' (Arceneaux 2005, 25; cf. Castells et al. 2007, 114-15; Goggin 2006, 107–14). Adam Burgess wrote in his study of the scare in North America and the United Kingdom that it was the result of a 'culture of fear.' He argued that the news media's focus on sensational issues, an obsession with personal and environmental health, and a cultural imperative towards consumer protection, colluded to stoke concern in Anglo-American societies. He posited that concern was linked to the 'invisible' and 'incalculable' sense of risk in contemporary societies. He compared the sense of risk to 'the childish fear of the dark' (Burgess 2004, 150–2).

Most studies on cell-phone use, risk, and safety have focused, not on health issues, but on surveillance. Researchers have often used the notion of 'perpetual contact' to describe how cell-phone use facilitates degrees of surveillance. Studies have made the case that perpetual contact through voice communication or text messaging was unwelcome when it made the cell phone into a 'digital leash' or 'digital panopticon' (Churchill and Wakeford 2002, Nafus and Tracey 2002, Puro 2002, Robbins and Turner 2002). Perpetual contact was welcome when it was agreed upon 'mutual monitoring' among family and friends (Green 2003, 33). Recently, with the shift to screening practices, studies have focused on how cell-phone users engage in mutual monitoring via 'pervasive photography' and 'intimate visual co-presence' (Gye 2007, Kato et al. 2005, Rivière 2005). Users maintain perpetual contact through taking and exchanging numerous photos of their everyday routes and routines. Also with such screening practices as pervasive photography, concern has been expressed about how cell phones might be used for illicit activities such as making 'up-skirt' videos or taking photos of copyrighted material (Castells et al. 2007, 118).

Concerns about cell-phone use, risk, and safety have been quite varied, but at the heart of the concerns were fears about what was visible and, therefore, known and what was invisible and, thus, unknown. In recent years, citizen journalism has made the third screen a mobile monitor for our fear. In order to interpret how citizen journalism extended and foreclosed certain ways of seeing cell-phone use, risk, and safety in our post 9/11 era, we must pay attention what was shown and what was not shown.

The Third Screen as Mobile Monitor

In citizen journalist reports of the London and Madrid transit bombings, dazed survivors were shown and corpses were visible in the usual way they are glimpsed partially in standard news reporting. In the aftermath of the tsunami in Asia in 2004, the 'brown bodies' of South Asians were shown but not the bodies of foreigners who also perished (Robinson and Robison 2006, 94). In citizen journalist reporting and standard coverage of the 9/11 attacks, 'images of corpses, body parts, and human gore were absent' (Zelizer 2002, 64). Citizen journalist and standard reporting offered a way of collectively 'bearing witness' through images in order to work through 'the difficulties that arise from traumatic experience' (Zelizer 2002, 52). Citizen journalists also used the third screen to insert themselves, figuratively, into historical events as 'participants and adjudicators' (Sobchack 1996, 7) in the way that bystander George Holliday used a camcorder to insert himself into history by recording video of Rodney King's assault.

In the images from Abu Ghraib Prison, American soldiers inserted themselves quite visibly into most of them, whether fully or partially represented by a gloved gesturing hand. The soldiers sometimes posed, grinning and with thumbs up, beside dead or frightened detainees. Their 'trophy' images were not meant for public viewing and bearing witness but, rather, only for fetishistic viewing by the perpetrators and by members of their tele-circle back in the United States to whom they sent copies (Razack 2005, 349, Robinson and Robison 2006, 91). The images were sent as though they were vacation photos, as a means of maintaining perpetual contact with loved ones, as if to say, 'We are safe and, look, we have subdued the risk to our safety.' When the images were broadcast, North Americans bore witness, not just to torture from months past, but to a historical fear of the racialized 'Other,' who in our post 9/11 era is also Muslim and not only black.

For some North Americans, the images from Abu Ghraib opened our 'third eye' (Rony 1996). We watched ourselves being watched as the images were broadcast and published worldwide. In analyses of the torture, North American scholars widely noted how much the making and sharing of images from the prison evoked the making and sharing of lynching postcards in the late nineteenth century, which nearly always showed victims after death (Apel 2004). The immobility of the victims contrasted vividly in the lynching imagery with the mobility of the living and gesturing spectators (Raiford 2003). Canadian scholar Sherene Razack also noted how much the images from Abu Ghraib resembled photos from the 1993 Somalia Affair, in which two members of the Canadian Airborne Regiment (CAR) took 'trophy photos' of the Somalian adolescent they tortured and murdered (Razack 2005). Like the early cameras used during lynching a century ago and in Somalia a decade before, the third screen mediated life and death at Abu Ghraib. The third screen became a mobile monitor for surveying who and what is most feared now.

Conclusion: Screening Cell-Phone Use

As a final example, the Abu Ghraib scandal showed why the third screen is an aptly named medium. The third screen became a cultural form in North America between 2003 and 2006 as we watched ourselves being watched while it was used to mediate citizen journalism, micro-movies, and mobile TV. By this, I do not mean to diminish the hegemonic position of North America in the world. I want to suggest that the third screen became a site of struggle over the making of culture and history in North America. It was used to both maintain and challenge perceptions and experiences of mobile communication in our post-9/11 era. By paying closer attention to the third screen, we may gain further insight into our times and culture.

NOTES

Thanks to the editors for inviting me to contribute to this collection. This chapter benefited from feedback provided by Kim Sawchuk and by its anonymous readers. An earlier draft of this chapter was presented during the 'Two Days of Canada Conference' at Brock University in November 2006.

1 For a detailed explanation of how mobile TV is broadcast, see Gow and Smith 2006 and Steinbock 2005.
2 It remains difficult to get statistics from wireless carriers regarding how much content is downloaded or streamed through their networks.

3 Following David Morley and Kevin Robins (1995), Mizuko Ito writes that the 'current Euro-American fascination with Japanese technoculture has deeper roots than the recent turn to the *keitai* Internet [cell phones with Internet access]. Invoking Japan as an alternative technologized modernity (or postmodernity) is nothing new. At least since the late 1970s, with rapid industrialization and emergence as an economic and electronics powerhouse, Japan has confounded Western models of modernization and technologization. Harking back to the international attention focused on Japanese management and electronics in the 1980s and 1990s, current Western interest in Japanese mobile phones and technoculture echoes a familiar mix of fascination and unease' (Ito 2005, 2).

4 The 1999 New Media Exemption Order made broadcasting that is accessed and delivered over the Internet to Canadian viewers exempt from regulation by CRTC.

5 Chris Anderson named this practice of digitally extending content in *The Long Tail: Why the Future of Business Is Selling Less of More* (2006).

6 For more studies on identity and culture, see Castells et al. 2007.

7 Intimate Strangers: The *Keitai* Culture of 'Belonging-without-being-with'

SANDRA BUCKLEY

The mobile phone or *keitai* is now the most widely used and familiar communication technology in Japan. Figures for *keitai* usage and phone-based Internet access for Japan outstrip even the hi-end phone users of Norway. Mobile phone usage is significantly lower in North America than in both Asia and Europe. Just 69 per cent of Americans own a mobile phone as compared with Japan, where mobile phone ownership and usage now exceeds land lines. Text messaging and e-mail traffic are low, with just 350 million wireless messages sent per month, in contrast to both Europe and Japan with an excess of 30 billion messages each per month. Twenty per cent of U.S. mobile phone subscribers utilize text messaging, compared to 50 per cent in Europe and 80 per cent in Japan.

Eighty per cent of the 80 million mobile phones in use in Japan today have Internet access. This access is linked to more than 60,000 Internet sites, and this number of access points increases daily. MobileMedia, a credible Japanese Internet-based research company, reports 58.7 million active mobile net users, with almost half of these users subscribed to one provider – the market dominant NTT and, more particularly, its subsidiary DoCoMo I-mode. I-mode simply refers to Internet-ready, while DoCoMo is a play on the Japanese expression 'dokomo,' which can mean both 'anywhere' and 'nowhere' depending on the context in which it is used. I want to return at the end to explore the playfulness and appropriateness of this linking of Internet-ready mobile communication technologies and the ambiguity of 'dokomo,' anywhere and nowhere. Perhaps in anticipation of the global potential of I-mode, NTT opted for an anglicized pronunciation for their branding strategy, transforming 'dokomo' into DoCoMo.

I-mode phones provide access to on-line shopping, banking, train schedules, dating sites, GPS maps and locator services, music stashes and voice mailboxes, and text and photo messaging. While a similar level of service access is increasingly available to mobile-phone subscribers in Europe and North America, what is distinctive in Japan is the level of active use of these *services*, or what might be described as a technological 'fluency' among Japanese *keitai* users. Of course, the 3G (third generation digital) mobile phones can also be used for a simple telephone call, but in Japan text messaging actually exceeds voice calling as the most popular communication mode. Animated phone-based text messages are a new phenomenon that is extending the user population further down into the youth market to capture increasing numbers of middle-high and even primary school students. The largest and still fastest growing user groups in the new mobile networks in Japan are teenage and young adult females. This marks a significant shift from the original target market of businessmen and travelling salespersons, as we shall see.

The *keitai* is an essential communication device in an urban environment characterized by an intensity of movement in the in-between spaces of commuter life. In everyday lives where 'not-at-home' is increasingly the normal condition, the *keitai* cannot simply be reduced to either its value as utility or accessory. The cultural practices surrounding the *keitai* are significantly transforming both spatial and interpersonal relations. It not only functions as a communication device but exceeds this technological functionality to become a strategic channel of emergent movements towards the renegotiation of the configuration of time, space, and memory – core elements of the architecture of sociality.

In this chapter, I will explore recent developments in the technology and its applications, as well as the sheer density of *keitai* usage, before stepping out into an exploration of the culture of the networked landscape of contemporary Japan. Much of what has been written about *keitai* and mobile phones elsewhere in the world treats the cultural practices of usage in isolation from design, production, marketing, and the rapidly expanding economics of telecommunications, despite the fact that these provide crucial context for understanding the rapid growth of *keitai* usage and the distinctive characteristics of *keitai* culture. In an attempt to begin to describe the new relational formation of intimate strangers within this emergent pervasive sociality, I have found that there is much value in recognizing the interplay between experimentation in the domain of the formal economy of the communications industry and the informal economy of sociality of the *keitai*. The two are arguably 'in play' in the new landscape of Just-in-Time (JIT) social relations, which I describe here as 'belonging-without-being-with.' In this chapter, what links these two, the formal and informal economies, are three street stories of *keitai* usage, each of which, in its own way, describes the processes at play between technology and culture when *keitai* take to the streets.

Innovative design of the hand-held units and related peripherals is an essential element of market-share capture for the Japanese telcos. This is an industry where technology lag time between competitors is staggeringly short, as 'catch-up' is essential to survival in the marketplace. In a pattern that runs counter to the American competitive market model there is a surprising level of sharing of core technology between competitors. Japan's telcos have selected hardware and software compatibility over barriers to competition. Nippon Telegraph and Telephone (NTT), in particular, has identified compatibility as a strategy for gaining, rather than losing, market share. And its smaller competitors also frequently share technology in a bid to gain joint advantage around a new service option not yet available from NTT. It is the special software features, accessories, and hardware design that finally differentiate in the Japanese market, where comparable high-quality technology is assumed as a given, leaving other variables such as ease of use, image branding, and options for individualization to drive market advantage.

Popular peripherals include everything from interchangeable dial plates to decorative covers, designer straps, ornaments, phone deco-stickers, and flashing or luminous antennas. Luxury packages can include anything from titanium shells to diamond-studded keypads, and on-line services such as a daily start-up message from one of an array of popular media stars or a personalized verbal or text-based daily calendar call detailing your commitments and any reminders such as birthdays, anniversaries, and work deadlines. Phones can be programmed to ring differently at different times of the day or in response to customized listings.

Since 2001, Japanese I-mode subscribers have been making extensive use of the music file transfer capacity of mobile phones, and this feature has added greatly to sales in the youth market, where phones have largely displaced the previously popular MP3 and mini-disc music entertainment platforms. Video and photo messaging have been available since late 2002 in Japan. These same functions became available in Europe and North America in many standard subscriber packages from late 2003, but the user uptake levels remain limited. Sha-mail (a term combining the *sha* of *shashin* – photo – and the mail of e-mail) is the name for the popular image messaging in Japan. Sha-mail is the latest innovation in Internet-ready phone services offering digital imaging messages as still shots or video clips which can be edited 'live' while making a phone call or text messaging. The dramatic entry of Japanese telcos into the realm of digital and video imaging has already hit hard in the retail sector, with camera sales plummeting and the once profitable film-processing industry facing major cutbacks. Electronic and camera specialty shops were selling video and digital cameras at discounts of 20 to 50 per cent in the early summer of 2004, as video-capable phones topped over 80 per cent of all new phone sales. Images can be printed at public user-pay digital photo printers now located in railroad stations and department stores.

An additional accessorized option on flip-top and slide phones is a display window showing a loop of personalized animation, still or video clip. These might be edited by the *keitai* owner for personal decoration on their phone, I-moded as a gift, or exchanged among friends. These image-loops can also run on the screen of an open-face phone when not in use. *Keitai-komi* (a wordplay based on the English 'communication' and 'community') and *I-furendo* (Internet friends), both terms for mini-networks, often share a common display image or video clip – a group-based practice to be returned to later.

Market economics and marketing strategies have been undeniable factors in the strikingly different take-up rates of mobile phones and text messaging in the North American and Japanese markets. Land-line phones remain significantly cheaper than mobile phones available from most U.S. service providers, and billing rates and calling cards all continue to favour land-line systems. This is in contrast to Japan, where waiting periods for a land-line phone can be as long as three months and the start-up fees are commonly in excess of $600. *Keitai* are immediately available, and start-up costs are low. In the United States and Canada, desktop computers and PCs have achieved a far greater level of domestic usage than in Japan, where space limitations in the home, lower average hours spent in the home, early domestic high pricing, and delays in approvals of patents all combined to keep household computer ownership as low as 40 per cent even by 2000, and most of that growth was directly related to the Internet. While the bulk of Internet access and e-mailing in North America takes place on desktops and PCs, in Japan the highest level of personal Internet traffic is generated by *keitai*. Japan's home computer sales stalled in the wake of the proliferation of affordable, flexible, wearable, Internet-ready mobile technologies.

In an attempt to compete with mobile hand-held devices, Japan's computer companies have focused on adapting designs from other environments such as the military, medicine, and the science laboratory in attempts to create mobile and miniature computer platforms. For example, a miniaturized screen mounted on a lightweight glasses frame, just to the edge of the primary line of vision, offers a hands-free digital display unit. This device, which was first developed to stream information to soldiers in the battlefield and to surgeons in the operating theatre, was featured in prototype at Japanese computer and multimedia trade fairs in 2001 and 2002, but it has yet to be commercialized. There is growing speculation that the Japanese computer industry is simply biding time while it observes the future of I-mode technology in phones before leaping into major new R&D investment. A multimedia computer platform that combines digital gaming, music, cinema, and television with Web access is anticipated to be the most likely direction in which the industry will move.

In this vision, the home computer would move out of the study and children's bedroom into the family room, with individual networked consuls in separate rooms as required. The PC would essentially become the FC (family computer) in the domestic setting. In this scenario, the computer industry would abandon Internet communications to the telcos. This new hybrid of home entertainment systems and fuzzy-logic home electronics could be adapted to interface with the *keitai* of the family members, allowing for remote digital access. A teenage daughter could dial-in from her *keitai* to her home to program the video to tape a television show she is not going to be back in time to watch; a housewife shopping in the supermarket could access the hot water system to trigger the bath heater for the family evening bath; and the family physician could monitor the heart rate and blood pressure of an elderly patient recuperating at home after surgery. Each of these functions is currently operational in Japan, but access to these services is limited to top-end subscribers, and only those who can afford a fully networked (fuzzy logic) home electronics system. Fuzzy logic is increasingly a standard feature of housing packages offered by Japanese real-estate developers. While William Gibson's recent novel *Pattern Recognition* (2003) features fuzzy logic functions similar to these as merely the stuff of futuristic sci-fi, they are in fact already everyday practice for some Japanese households.

This level of networked access requires extensive collaboration across industries and a high level of compatibility. Incompatibility has been a major obstacle to the growth of new communication and information technologies in North America. Technological barriers between service providers have been a key strategy in the fight for market share. AT&T launched two-way messaging in 2000, but its service was limited to AT&T subscribers for the first two years. Telcos in the United States focused on limiting the risks of subscriber poaching or brand switching by building market-share protection into their systems and creating high hardware costs with discounts tied to brand loyalty. This is a typical example of a U.S. market model of competitiveness that actively limits consumer access to new technologies in the interest of protecting corporate market share. By contrast, DoCoMo in Japan determined this market model to be archaic and unsuited to the speed of product diversification and redundancy in the field of information and communication technologies. New software packages have now freed up access between some systems to facilitate text messaging in the United States and Canada; however, there remain many glitches between carriers.

North American telcos also remain reluctant to invest deeply in SMS (Short Message Services), mainly because of the pricing structure for billing. The United States and Canada adopted the Japanese and Korean strategy of volume-based pricing instead of the pay-for-time model that still dominates the North American voice-call market. At anywhere from $5–$10 per 100 messages, SMS is substantially cheaper than pay-for-time calls. Market analysts argue that the volume-based pricing packages of SMS text messaging have limited the profit potential for this technology, and that U.S. carriers have been reluctant to sink funds into R&D and market development primarily for this reason. Having followed the Japanese lead in volume-based pricing, telcos are now unable to reframe this to achieve higher returns. When one major Korean telco attempted to shift from cheap volume pricing, there were street demonstrations by thousands of young subscribers.

Why is it that DoCoMo has become monstrously profitable using the same volume-based pricing structure that apparently limits profits here? The answer lies in the marketing strategies adopted by the Japanese telcos and their fundamental departure from some of the strongest operating principles underpinning U.S. competitive markets – product protection, market monopoly, and barriers to entry. A couple of U.S. carriers, AT&T and Verizon, have decided to focus on promoting the text-messaging market and have established SMS-linked websites where they offer on-line support for the creation of SMS 'comnets' (community nets) around new release movies or celebrity sightings.

Text messaging is growing in popularity, with contracts increasing across the youth market in the United States and Canada at 20 per cent per year in 2002 and 2003, but easy access to the Internet remains more of a myth than a reality. Unlike the Japanese telcos, North American carriers continue to insist on controlling much of their customer Internet access and content development, functioning more as gatekeepers to the Web than as gateways. Most North American systems still use pay-for-time billing for Internet access other than text messaging, and also charge an additional access fee per website dial-up. In this approach, the mobile phone functions simply as a dial-up tool.

Web content providers in North America are charged high fees for the privilege of 'nesting' on a telco portal. The model is based on television and other media advertising contracts – with the product developer buying time or space for consumer access. By contrast, DoCoMo provides an interface to over 60,000 Web servers, either through an I-mode converter operated by DoCoMo or through a corporate converter owned and operated by the Web content provider. DoCoMo technicians provide I-mode conversion support to any potential content provider and actively encourage developers of new I-mode compatible content. In other words, DoCoMo has opted to offer technology access both to its I-mode converters and its in-house simplified iHTML software, rather than attempt to control access through a single DoCoMo portal, and has adopted a strategy of alliancing towards new product developers. A perfect example of this was DoCoMo's support for the development of downloadable ring tones and *emoji* (creative graphic notations used as a fast iconic shorthand for text messaging). DoCoMo made the technological adjustments to accommodate a wider range of ring tones and supported the software and interface development of *emoji* provider sites. They also backed the launch of software that would allow users to create, utilize, and trade individual icons. These two highly popular elements of the personalization of the *keitai* not only added new dimensions to the *keitai* culture but also created a simple and enjoyable pathway for millions of young Internet-shy Japanese to learn the basic skills of Internet access and information downloading as a first step into this new multimedia Internet world of communication and consumption.

I-mode offers a navigational design that creates the impression of transparent Internet access for the user and maximum support for content providers. DoCoMo states that its goal is to offer a seamless vision of wireless and Internet technology. Unlike in North America, mobile platforms are seen as potentially displacing the PC in Japan for the majority of daily Internet interactions for most individuals over the next several years. The applications for desktop and PC are expected to become increasingly limited to the corporate and institutional environment.

So how does DoCoMo make money through its Internet services? DoCoMo's marketing and R&D teams have jointly created a new profit strategy in the communications industry. DoCoMo, and its parent company NTT, have cultivated a grey zone between banking and finance and communications by grounding their I-mode profits in a 9 per cent share of all billings by Web content providers accessed through I-mode phones. The charges for services and goods purchased by the I-mode link are added to the monthly phone account. DoCoMo is thus able to control its 9 per cent share and has accessed an entirely new area of business through this step into the finance and credit markets. By the end of 2004, subway commuters and bullet train and airline passengers were all able to book and purchase fares using their *keitai*. Several major department stores and supermarkets are also in the process of developing on-line payment for purchases to replace in-store payment at checkouts; this will function in a similar mode to direct debit but require no debit card. Charges for on-line information services are very low (equivalent of just a few cents), but DoCoMo is strategizing on volume not high pricing. Access to DoCoMo on-line sales is now seen as a key to competitiveness across commercial market sectors. Financial analysts in Japan are predicting that the future of I-mode fiscal and credit management may spell the demise of the current banking system and result in a major realignment of industries that could see corporations such as Sony and DoCoMo stepping into a leading role in a reinvented financial sector. Such a shift should not be a surprise in a global economy where the emphasis is increasingly on the flow of information over goods, and where the virtual profits of futures and venture capital can make or break financial markets and investors.

Technology and market strategies have been discussed; now, strategies of everyday use of *keitai* culture need to be addressed. What DoCoMo is describing as seamless wireless and Internet access is what is widely known today as pervasive computing – minimum effort for maximum access to digital communication, information, and tasking. Japan's *keitai* is the most accessible current platform of pervasive computing and has extended access to a new level of mobility and range of audience or reception (in both senses of that word).

The first mobile messaging device in Japan was the POKEBERU (pocket bell), and it was launched by NTT (DoCoMo's parent company) in 1968. The pokeberu design was corporate driven and was closely linked to the role of subcontracting in achieving the cost savings generated by the just-in-time (JIT) delivery systems often cited as a key factor in the so-called Japanese economic miracle. Rapid communication was essential to JIT delivery between suppliers, buyers, production lines, distributors, and merchandisers. The pokeberu was the prototype of the text messaging system still in wide use internationally, and known as the 'beeper' or 'pager.' The pokeberu signalled the carriers that they needed to call back a designated number or call into a message centre. This technology was taken up early in the 1970s by harbour tugs and ferry services to facilitate rapid information access while at sea, such as weather and channel traffic conditions. In the late 1970s, it was adopted by the railroads, a popular means of transport for salesmen and other travelling corporate representatives, to facilitate communication during commuter hours, thus avoiding downtime while in transit. In 1987, NTT launched the first individualized mobile telephone. Far larger and heavier than contemporary palm-sized lightweight *keitai*, it was still a major breakthrough and rapidly became popular among businessmen and travelling sales staff.

From the outset, the *keitai* evidenced particular usage patterns that distinguished it from the culture of land-line telephones. The family phone was traditionally located in the entrance to the home, symbolically mediating between the inside and outside of the family space. While over time the home phone has migrated into the kitchen and family area, it is still often 'dressed' in a lace, embroidered, or otherwise decorative cover in a gesture of 'domestication' of this technological interface with the unsolicited intrusions of the external world. While there are social and linguistic protocols for anyone walking past a ringing phone in an office or home to pick up the receiver and take a message, the *keitai* has always been perceived as an intensely personalized device that no one else would usually handle. Despite the fact that on most Japanese mobile phone plans a local call would cost nothing extra, there is a social taboo against asking to borrow a friend's or stranger's mobile. A ringing *keitai* will be left to ring in the absence of the owner. While this reluctance to handle anyone else's *keitai* may reflect issues of cleanliness and hygiene, major concerns in Japanese society, there is certainly also a link to the highly individualized nature of *keitai* ownership and use.

The personalization of the *keitai* includes even the pattern of exchange of phone numbers. Most Japanese *keitai* owners still only offer their mobile number to immediate family, an inner circle of friends, and key work colleagues or clients. Many adult Japanese maintain two *keitai*, one personal and one for work. A first tier of contacts is given the appropriate work or personal *keitai* number. A second tier is offered e-mail or *keitai* (remember that e-mail are accessed by *keitai* and offer more immediate or continuous access than a land-line phone), and a third tier receives only access to office or home land-line. The process of selection or screening is now further refined with the addition of call editing as a standard *keitai* feature. A mobile can be programmed to automatically edit out calls from specific numbers by redirecting to the message bank or a 'no answer' mode that rings out. Phones can also be programmed to allow a call waiting signal for some selected and prioritized numbers.

The level of personalization is extended to the physical appearance of the phone in Japan more than in any other country. While it is not uncommon today to purchase interchangeable phone 'dial plates' that can be switched for different contexts or even to colour coordinate with what you wear, in Japan an entire market has evolved for *keitai* peripherals or ornamentation. Specialty stores sell everything from crocheted to titanium faceplates, a myriad array of decorative straps, carry cases, and, of course, thousands of ring tones. Individuals personalize their phones down to the last detail, including animation or video clips that are visible on an external display window. The extent to which the *keitai* is embedded in the everyday life of an individual and their public performance of self elevates it to the level of a hybrid of fashion, communication device, and prosthetic.

The *keitai* is a wearable technology now incorporated into the design of jackets, belts, handbags, and briefcases for fast and ready access. The most common carrying technique for men in Japan is a fashionable, often designer brand, strap that is hung around the neck or linked to the belt and pocketed. Women are more likely to carry a phone in a purpose-designed external pocket of a handbag or dangling from the wrist on a short strap decorated with any number of beads, and miniaturized figures of favourite anime or television characters. More often than not, though, when on the move, the *keitai* is carried in the palm of the hand. This is partly due to the generally respected social taboo on ring tones in closed spaces like offices, trains, and restaurants, which leaves the user reliant on the vibrate mode to recognize an incoming call or text message. It is hard, however, not to also interpret this sustained hand to peripheral contact as the bodily expression of the *keitai* as a prosthetic extension, an articulated communication device, embedded in the movement of the body if not yet a literal implant.

Like its predecessor, the pokeberu, the DoCoMo I-mode phone is also about JIT access. In the 1990s increasing levels of corporate and institutional resources were concentrated on centralized, digital archiving of information across Japanese industry sectors. However, a number of key productivity and efficiency surveys demonstrated that by the mid-90s less than 20 per cent of the information individual employees needed to achieve their daily tasks, from factory production line to sales teams to executive boardrooms, was actually stored on corporate mainframes. The majority of information was accessed, and could only be accessed, through contact with other individuals. A major drive behind the development of the *keitai* technology in Japan was an awareness at the highest levels of Japanese industry that a technology that could facilitate remote mobile communication would play a crucial role in enhanced efficiencies. The earliest advertising images of the *keitai* user were of suited businessmen travelling on the bullet train, rushing through an airport, or sitting in a meeting room accessing vital information back at their head office. In the corporate environment, the *keitai* remains one element of a communication ecology that also includes palm pilots, PCs, beepers or pagers, text messaging, e-mails, voice messaging, and, in some cases, Limited LANs or Wi-Fi networks. For the corporate user, the *keitai* is a tool for linking place-to-place and bridging geographic barriers to face-to-face communication. It fills the communication gaps and information lags of PCs, mainframes, and land lines. The primary goal of the corporate *keitai* user is the rapid accessing of people for the purpose of sharing essential information.

However, the largest user group of *keitai* in Japan is no longer businessmen, but rather female and male teenagers and young adult women. The *keitai* remains an essential tool for accessing information, but the practices of JIT access have been transferred into the domain of youth culture, where there has emerged a JIT model of social relations. This is an economy of sociality where value is measured in degrees of accessibility, loyalty, responsiveness, and reputations. Familiarity and information accumulate and circulate as the capital of this space of pervasive sociality.

Let me describe briefly three separate scenes I observed in Tokyo on a recent visit and then step further into an exploration of this pervasive sociality.

Scene 1

A young woman stands outside Harajuku Station, a main commuter window onto a neighbourhood that is one of Tokyo's centres of urban youth culture and therefore, by definition, a major consumer district for brand designs, knock-offs, and fringe fashion. She is dressed in a style that clearly indicates that she is a member of a *sutureeto* (street) clan or tribe. Holding her *keitai* in her hand, she is watching the passing crowd pouring in and out of the station as she thumb types messages into the keypad of her mobile. This is a member of the so-called thumb tribe of Japanese youth. The technique of holding the phone in one palm or between the palms of both hands and thumb typing is the most common form of usage of *keitai*. It is far less common to see a phone held to the ear for a voice call. She is 'talking' with a friend who has come out of the wrong exit at the station and cannot find her. She suddenly waves and takes off in the direction of the person waving back to her across the street.

Scene 2

A 'Peace in Iraq' demonstration is announced for the weekend in downtown Tokyo. The main organizers are an alliance of environmental groups that have selected Earth Day for this peace march. A smaller alliance of more militant anti-U.S. and anti-war protest groups wish to crash the demonstration march and are strategizing to bypass official permits and disrupt the officially sanctioned march. An hour before the scheduled start of the march, these alternative protestors begin an intensive net of mobile phoning and text messaging. There is no determined strategy for 'crashing' the demonstration. They ride the subway individually and in groups of two or three moving into the general area of the official protest route. As they travel, they begin to text message one another using a designated temporary website as their virtual meeting place. This is a weblog utilizing blogging software in combination with video clips. As heavier patterns of convergence become clear on the GPS-based tracking system, the patterns and density of movement of protestors shift towards the trajectories of the highest concentrations of messaging. Those who recognize that they are operating at the margins of an emerging pattern of convergence voluntarily stop messaging and follow the nearest emergent line of movement. There is no attempt to compete or redirect. Eventually the protestors come up to street level from the *chika* (underground) at a variety of locations and march loudly and visibly along non-approved routes, cutting across and disrupting the collaboration of police and protestors in the main march and capturing, even briefly, the attention of the media.

Scene 3

A group of several teenage girls stand queued outside one of the most trendy fast noodle counters in Harajuku. They stand talking to one another about their plans for the day and whether or not they should stay in line for what could be another half hour or so of waiting. Each of them is thumbing her *keitai* as she speaks with her friends. I ask if I can photograph them, and they happily start to describe their separate *keitai* connections. Two of the five are staying linked to friends who are in other parts of Tokyo, one at a wedding and the other shopping in the other most popular shopping district, Shibuya. Two others are linked to local Harajuku LANs (local area networks – unlicensed, wireless, broadband networks).

The LANS they are tapping into are run by alliances of retailers, community groups, entertainment outlets, and local government promotional agencies. You can go on-line to one of these local networks carrying constantly updated information on what is happening in the area, from discounted fashion to concerts and unplanned street events. Or, your I-mode connection may instantly channel you to the nearest LAN content using GPS technology to locate your handset. Or, you may have bought an I-mode package that offers discounted pricing for text and voice messaging in exchange for advertising and LANs access. At any point in time, your *keitai* screen will stream updated information about events, sales, entertainment, and restaurants in the Tokyo city ward you are currently moving through. This information can be tailored according to your 'reputation,' an accumulative profile of individual download and consumer patterns.

The five young women maintain a continuous flow of conversation, sharing the information coming in from their *keitai* as they try to decide if they should linger in the queue or move on. Their thumbs remain in perpetual motion, and their eyes shift from screen to friends and back. They maintain a lightweight level of e-chat with other net *furendo*, messaging a flow of continuous trivial commentary on the ordinary. The practice is sometimes described as grooming – a phatic mode of communication intended as a gift or signal of attention or affection, an emotional gesture, rather than an information-driven message. This is essentially the opposite of picking up the phone to call a friend with news, the new, the unexpected, the extraordinary. Text and icon snapshots of the mundane, the 'not-new.' Detail is not offered, just snippets or fragments: 'just in time delivery' of just enough information. This low level e-chatter was described by one young person as 'e-noise.' The static of the everyday, just present on the horizon of the here and now, resonating but not intruding. It is a non-present presence. Omnipresence without proximity. An intimacy among strangers.

The one-to-one text messaging of the girl rendezvousing with her friend outside Harajuku Station represents the simplest of the common functions of *keitai* in youth culture as an instrument of coordination. In this usage, the *keitai* facilitates and coordinates movement towards a common location, a meeting place. What is new about this? The phone has long been a tool for planning. What has changed is the shift from setting a meeting time and place to a new dynamic whereby two or more people move towards a potential coming together through an unfolding process of identifying and eliminating options via text message updates. Two friends might begin an e-chat early in the afternoon as they both leave their respective schools, commute to their cram school tutorial, grab a bite to eat, move on to a piano lesson, and plan their homework schedule. Each always knows where the other is, and they circle around the possibility of meeting.

This non-present presence displaces notions of lateness with a not-yetness, a perpetual state of anticipation that is pleasurable in itself. Two people are not yet in the same place but are text messaging one another as they clarify how far away from each other they are and their respective routes: on which subway, at which stop, on which corner, crossing which street. Maybe they will adjust their movements to achieve a more efficient or less crowded, or cheaper pathway to a common point of intersection. In response to a text message announcing an unexpected event, an i-ad for a bargain sale, or another convergence of friends messaging elsewhere, they may divert and redirect their movements together or separately. They may meet or they may not but remain linked, one on one or one to many.

Immediacy of access and response is the expectation between *I-furendo*. Being in contact, in touch, promotes a sense of perpetual planning around the potential of meeting, but there is no compulsion to meet. The very act of being in touch often finally substitutes for a face-to-face encounter. The five young girls standing outside the noodle counter are deciding if they will physically meet up with the friend shopping in Shibuya. The decision to stay in line at the noodle counter is not a decision not to join her, but rather a decision not to be with her – belonging-without-being-with. They continue to e-chat just enough information enough of the time to stay linked. They offer her shopping advice when they receive I-moded photo images of the various shoes she is trying on. They print out the images for better clarity on a hand-held printer one of them pulls from her backpack. They text message news of a super sale at a popular local shoe store that has just posted an i-ad on the Harajuku LANs-net they are accessing. Their other friend I-modes video clip of the bride throwing her bouquet at the wedding she is attending, and the five girls all cheer and clap.

Keitai contact is for many young Japanese a suspended event that sustains a condition of belonging-without-being-with. This condition is in no way experienced as either an absence or a longing for. I am suggesting that belonging-without-being-with is the emergent condition of community formation fostered by *keitai* technology. These relationships are experienced through a connectivity that is person-to-person rather than place-to-place or face-to-face. This sociality exceeds physical contact or physical space. These five girls are literally 'dokomo' – everywhere and nowhere, living in a layered present of layered presences. Time and place are spatialized outside the familiar geo-chrono axis and experienced instead as a relational field of pervasive sociality. The present and presence are no longer bound to the necessity of 'being there' in a relationality of belonging-without-being-with. Distance melts into immediacy in the realm of itinerant intimacies.

One Japanese researcher has described this blurring of previously clearly demarcated spaces and the traditional spatialization of relations as a reconfiguration of power geometries. Mizuko Ito (2005) has undertaken a number of *keitai* usage pattern surveys and interviews among Tokyo high-school and university students. She has concluded that Japanese youth utilize *keitai* communications to subvert the power of both parents and teachers to monitor and control their communications. At home, children can maintain contact with friends via the *keitai* without the mediation of parents, whom once controlled access to the centrally located home phone. It is harder for parents to monitor whom the children are in contact with or the level of contact. Many schools ban *keitai* use during school hours, but text messaging makes this a difficult policy to administer. Similarly, many companies discourage, even if they can't ban, private *keitai* usage during work hours, and this is equally difficult to enforce.

Ito and others who have recently considered the practices surrounding *keitai* usage in Japan have tended to limit their discussion to questions of a renegotiation of power in specific institutional spaces that house certain power relations – household and family, school and teacher/student, workplace and employer/employee. However, I believe that the impact of *keitai* culture exceeds this, bringing far more into question. A female graduate student is sitting in a café with her professor and fellow students, and her boyfriend calls; a businessman is walking through an airport with his boss, and his child's teacher calls; a working mother is sitting in a meeting with her staff, and her broker calls … in that instant of answering the call, each individual's role switches. An observer can track a significant shift in both verbal and non-verbal communication, and there is a temporary disconnection from one role into the time/place configuration of another role. This remains true when text messaging replaces the voice call. Although the role switching is no longer observable, the effects remain. As the businessman talks with his boss about the negotiation they are heading into, he is simultaneously iconing a comforting message to his child that he will be home tonight.

Text messaging has dramatically increased the usage level of mobiles, as individuals feel that they can less obviously or intrusively stay connected to their multiple networks and effectively manage opaque role switching. In Japan, where social taboos and protocols abound, the level of *keitai* usage in public spaces was severely limited in the years prior to text messaging due to the social impropriety of both the phone ring itself and the act of imposing your end of a conversation on total strangers. The vibrate mode and text messaging facilitate non-disruptive pervasive sociality. Within *keitai* culture, multiple connectivity is not experienced by an individual or their peers as either intrusive or disruptive. There is an assumption of non-competitiveness. Connectivity is not exclusive. There is no expectation within *keitai* culture of exclusivity. The notion of having someone's full attention has been diverted to having open access.

Some Japanese sociologists are already decrying the impact of *keitai* culture, citing risks of low attention levels, a dissipation of relationships, and a levelling of emotional interaction (Cooper-Chen 1997, J-Wave Editorial Group 1993, Kojima and Yoshiaki 1998). This, however, reflects a long history of Japanese researchers associating new technology and popular culture with aberrant or anti-social behaviours. Video became synonymous with pornography, video games with sexual violence, anime with the intensely private and obsessive behaviours of the otaku. Bizarre and isolated criminal cases are used to evidence and popularize simplistic cause-and-effect arguments. The media is already filled with editorials on the breakdown of family relationships and the threat posed to the authority (i.e., institutionalized power) of parents and teachers when children can create networks of relations beyond the reach or control of the home and school. What answers do we find if we examine the popularity and pervasiveness of *keitai* culture, not from the premise that the impact of new technology must be negative, but rather starting with the simple question of 'what is the appeal of *keitai* culture?'

Japan is a society where daily life has been intensely compartmentalized into dichotomous spaces: home / not at home, work/leisure, public/private, ingroup/outgroup (*uchi/soto* – in Japanese), family/nonfamily, cohort/non-cohort. Discrete regimes of appropriate practices and terms and conditions of relations are rigidly defined within each of these bounded spaces. A density of honorific language and coded behaviours perform this compartmentalization and hierarchization into the everyday. The blurring of behaviours across bounded spaces carries a high social price tag. As *keitai* continue to enhance the potential for pervasive connectivity, dense networks of relationality perpetually exceed the space any individual inhabits physically at any given moment. Complex layers of co-present realities and the related processes of role switching refuse the familiar reductive dichotomies. *Keitai* culture enables people to experience the complexity of their lives simultaneously rather than sequentially. *Keitai* have created an invasion of privacies into the public and work spaces at the same time that this technology also supports the corporately driven notion of 24/7 access that collapses the home/work and work/leisure distinctions. Young women and teenage girls make up the largest, and still fastest growing, *keitai* user groups. In Japanese society, these two groups, teenage girls and young adult women, arguably experience the most intensely mediated conditions of identity and the most limited possibilities of group mobility or alternative lifecycle pathways. That they are the most likely to step off-track to participate on-line in the unmediated potentialities and mobility of belonging of *keitai* culture should not be unexpected.

Younger Japanese citizens are stepping out of their compartmentalized lives and negotiating wider networks of relations than have been possible under the previous restraints of traditional institutionalized authority and rigid disciplinary regimes both at home and school. For both students and commuter workers, *keitai* culture is reconfiguring the spaces of the in-between that had played such a crucial role as essential opportunities to temporarily exit the tightly managed circuits of everyday life. The most common spaces of the in-between have been the subway (*chika*), *kissa* (cafés), fast-food outlets, and station plazas and stairwells. The in-between has not been a static space of waiting or killing time, or dead time, as it is sometimes misinterpreted by sociologists and anthropologists researching Japanese popular culture. It is not a space that is inhabited or occupied. It is a space of movement that is defined by strategies of exiting the prescribed circuits of productive movement to engage in activities and relations that create alternative flows of sociality that exceed the limited or bounded possibilities of the everyday. It is no surprise, then, that *keitai* culture has taken to the streets, the subways, the stairwells, the cafés, and plazas at the same time that it has itself become the latest spatialization of the in-between.

Sutoreeto is a word widely used in Japan, a japanization of the English 'street.' Japanese youth culture is exemplified by the notion of *sutoreeto*. It describes, not a physical space, but a diversity of practices defined by fluidity and movement and deeply rooted in consumerism. While *sutoreeto* is intricately implicated in the mainstream circulation of currencies of desire, goods, and money, it is also a space of the in-between where the young *sutoreeto* protagonists write their own scripts. The emergence of a district as a known *sutoreeto* site will be marked by the availability of maps that indicate both the physical streets and the cultural and consumer attractions of the district. These used to be available as handouts at subway stations but are now also offered on-line and via LANs for *keitai* users moving through the district. *Snappu* (photo shoots) become a common scene in a *sutoreeto* district, and the area has 'made it' when it has a dedicated page in any of the encyclopedic *sutoreeto* consumer magazines. These magazines will offer not only a detailed description of what is the latest *sutoreeto* fashion, corner by corner, but also consumer outlets for each 'look' and pricing information. However embedded *sutoreeto* culture may be in mainstream consumerism, what defines *sutoreeto* is its volatility and fluidity. The magazines and *snappu* are constantly attempting to keep pace with and capture the shifting ground of the performance of *sutoreeto*. It is a perpetual process of deterritorializing and reterritorializing. *Sutoreeto* is always both radical and reactionary space. It is an unending negotiation of differentiation and integration.

Let us return now to the scene of the street demonstration I described above. The technology of *keitai* has been widely associated with the practice of swarming. From the anti-government movement that brought a regime change to the Philippines in 2001, to WTO and G8 protests since 2000, and rave concerts, swarming or flocking enables large groups to move to a single location without prior planning or notification and without a hierarchical organization requiring leadership. Through GPS and voice and text messaging systems, large numbers of individuals track emergent patterns of convergence and elect to cluster rather than compete through cooperative strategies of alliancing. It is interesting to note that the strategies DoCoMo is pursuing with its content providers to promote the growth of I-mode can also be characterized as more a process of alliancing for joint or shared gain than the usual market-driven model of competition and elimination. As in nature, swarming is characterized by a low level of rules or mandatory behaviours and a reliance on voluntarily subordinating individual preference to group dynamics. The underlying assumption of swarming is that there are always others more familiar or more informed and that the accessing and sharing of information across a group will lead to a stronger strategy than any individual in the group can achieve alone. Swarming is always an emergent and unpredictable behaviour. It is by definition a leaderless movement precipitated by a shared goal. It is also always transitory, and the swarm dissolves once a shared goal has been achieved or a destination reached.

In on-line and *keitai* culture, similar principles operate in weblogging or knowledge-blogging environments. Networks of *I-furendo* blog information to a common site and then disseminate this information either unrestrictedly or to a closed network. What is created can be described as an 'augmented reality.' From alternative independent news sites to political organizations (even ones as conservative as the ruling Liberal Democratic Party of Japan) or the latest *sutoreeto* scene, augmented reality sites are available to anyone who dials up. They have access to the accumulated experience and knowledge of all the net members, and as they participate, they also enhance and expand the specific field of cultural capital of the site and its network.

Crucial questions emerge around the emergent everyday practices of *keitai* culture. As the flow of information and relations is freed from familiar configurations or compartmentalizations of time and place as space – the home, the office, the bank, the classroom – the built spaces of the city are reduced to fulfilling only the residual functions that require people to be in the same place for any length of time or at the same time. The number of functions that require co-temporal proximity are being reduced as the functionality of mobile technologies expands. At the same, the relations of power that support discreet structures of power, or structures of discretionary power, are also potentially up for renegotiation. *Keitai* culture is characterized in its many practices by the condition of belonging-without-being-with. *Keitai* culture unsettles normative mechanisms of group and community formation and identification. I-moders do not settle, they roam across a multiplicity of co-present communities of practice and identifications living anywhere and nowhere. I-moders are not-at-home everywhere. One young Japanese woman when describing her *keitai* said to me, 'This is better than virtual reality. I can be everywhere all of the time.' What better description of the I-mode world of DoCoMo – or dokomo – anywhere and nowhere.

8 Terminal City? Art, Information, and the Augmenting of Vancouver

DARIN BARNEY

Art and love can only find their fulfillment in a vision of nature in opposition to our freedom.

— George Grant, *Philosophy in the Mass Age*, 102

I

Every city has its virtue. It has been said of Vancouver that its virtue is not so much civic as natural: that its excellence is attributable not to anything built or done by those who inhabit it but, instead, to the majesty of its situation in the centre of a rainforest, at the foot of the Coastal Range, where the mighty Fraser empties into an endless ocean. This is only half the story. The virtue of Vancouver is that it is a place where artifice and nature collide and, in that collision, bring something true forward into beauty. In this sense, in its very materiality as a city, Vancouver accomplishes the work of art more closely than does any other Canadian metropolis. The city's situation affords this possibility, but it has never guaranteed it.

A good place to witness this collision is (or has been) the University of British Columbia's Museum of Anthropology (MOA). Built by Vancouver-native architect Arthur Erickson, the MOA is made of concrete and glass and light on a design that evokes the post-and-beam style characteristic of the architecture of the first peoples of the northwest Pacific coast. It is tucked into a small forest on the ancestral land of the Musqueam people, its back to the city and its face to Asia. Entry to the building is through two, red cedar K'san doors carved by four Gitxsan masters (Walter Harris, Earl Muldoe, Art Sterritt, and Vernon

Stephens), which convey as much by their sheer weight as they do in their narrative of the first people of the Skeena River region. Very quickly the visitor is drawn by light through an opening onto a gentle downward ramp where she is surrounded by everyday things – Coast Salish bent-boxes and feast bowls, Kwakwaka'wakw house posts, a blanket woven by contemporary Musqueam artists Debra and Robyn Sparrow, and fragments of great Haida poles. The experience of this threshold is one of profound and liberating disorientation. Gravity pulls the visitor forward. The Great Hall of the museum is a massive, quiet space curtained by a fifteen-metre-high wall of glass, filled with the si-lent testimony of the things of the first peoples: house-posts, totem poles, massive carved creatures. Beyond the glass curtain, on the cliffs of Point Grey, amidst a carefully designed landscape of indigenous plants and grasses, stand two Haida houses, ten majestic totem poles from the Gitxsan, Nisga'a, Oweekeno, and other first nations, two carved house-posts, and two welcome figures carved by contemporary aboriginal art-ists. And, beyond all of these things, the Coastal Range as its turns northward, the sea, and Japan and China unseen in the distance.

It is difficult to describe the experience of this space and the things in it that open a world. Time moves slowly but surely here. It is a clearing for judgment, in which the virtue of the city of Vancouver stands light-ed, at the bottom of the ramp in the Great Hall of this museum. It is not comprised of the city's natural situation, its aboriginal past, its modern architecture, or its Asian futures, but the collision of all of these, a colli-sion that clears a space in which something true about the place can be registered in its astonishing beauty.

That a built space filled with things can make such a registration pos-sible testifies to the possibility of art. Lately, this registration and the pos-sibility it raises have been interrupted by the presence of a new wireless technology designed to augment visitors' experience of the reality of the place and its things. The technology, developed by a company called Ubiquity Interactive in cooperation with the MOA, the CBC, Telefilm Canada, and the Canadian Museum of Civilization, is a hand-held, multimedia, interpretive aid known as the VUEguide. The device meas-ures eight centimetres by fifteen, with a screen about half that size whose features can be activated by touching pixilated buttons using a small, plastic pen. The device also comes with a single earpiece connected by a wire. It resembles in kind, if not in elegance, any number of the small, portable, screen-and-earphone devices that have become customary ap-paratuses for inhabitation of urban, networked spaces. Inside the device

is a chip on which is stored data that is activated by infrared beacons located throughout the museum, prompting the viewer with a menu of
choices for access to additional troves of information.

Looking downward onto the screen of the VUEguide, pen in hand,
one enters a different space, a different world. It is a space of flows, a
world of digitized information, where time moves quickly and not at
all (Castells 1996). It is a brilliant world, to be sure, and its bias is towards overcoming the disorientation one might otherwise feel standing in such an unusual place. Plugged into VUEguide, one not only has
access to records tagged to specific objects in the museum, but also to
animation, audio narration, graphics, historical and re-enacted video
footage, and maps. One hears the voice of Haida artist Bill Reid reflect-
ing on the making and meaning of a frontal house pole. There is a love-
ly animation of the mysterious manner of making bent-boxes from a
single piece of wood. A model of Sea Lion House, as it stood at Quatsino
Sound around 1906, is generated on screen in three dimensions from the
perspective of the viewer standing in the museum beside a real arch-
way, bench, and house-post recovered from the site. An audio-visual
presentation shows the making of *Lootas* – a fifteen-metre Haida war
canoe carved by Reid and his apprentices from a single cedar log – pad-
dled between Skidgate and Vancouver for the 1986 World Exposition.
Walking around Reid's sacred sculpture, *The Raven and the First Men*, an
image on the screen rotates to match the viewer's perspective, and a tap
on the image of one or another of the piece's many figures activates
precise details of its symbolism and place in the whole. Reams of text
and images untagged to specific objects provide comprehensive information on aboriginal cultures, traditions, history, social structure, and
artistry. The material is uniformly rich and crafted with great care and
intelligence, deep respect, and attention to legibility and detail. The
VUEguide is a magnificent technological achievement, especially in the
context of a museum whose mission is equal parts cultural, educational,
and scientific.

How does this wonderful device, in augmenting the reality of the
museum, interrupt the experience of the collision singled out above as
the museum's particular excellence? There can be no objection on
democratic grounds to overlaying the space of flows and information
onto the world presenced by the Great Hall. Patrons had made clear
what the old space of the museum and its things left them wanting:
more information. Approval ratings of 85 per cent demonstrate that
VUEguide has been a whopping success in this respect (Ubiquity

Interactive 2006a, 13). Nor are the devices necessarily atomizing or anti-social. The non-immersive single earpiece allows for conversation with others and eavesdropping; the expertise accessible onscreen enables, and even encourages, spontaneous acts of popular education, as when a stranger corrects a neighbour's mistaken impression that the cracks in Reid's *Raven* are damage (VUEguide informs that these result from nat-urally occurring and self-correcting expansions and contractions of the yellow cedar). And, despite its wireless portability, the device is strictly situated, insofar as it functions and is meaningful only within the physical confines of the museum, in range of the infrared beacons that activate it. Finally, there can be no romantic appeal here to a 'pure,' non-technological, immediate experience corrupted by technological medi-ation: the Great Hall is, to be sure, always-already enabled by technology and mediation.

Nevertheless, there may be a difference between the world opened by the Great Hall and its things, and the world opened by the VUEguide, and between the ways in which access to these worlds is mediated and experienced. The manner in which a visitor moves into the world opened by the Great Hall is described above. With VUEguide, one does not move into this world as a visitor, but instead accesses a network of digitized information in the manner of a user, a label that describes anyone whose life practices are mediated by devices designed to ac-complish instrumental purposes. Stepping over the threshold onto the ramp, the user is immediately prompted by a signal that there is infor-mation available via the network, and so unfolds the *habitus* character-istic of beings that belong to the world of hand-held, portable, wirelessly networked information appliances. Eyes cast downward to the screen, hands ease into point-and-click dexterity, ears tune to the tiny speakers that cram them. As the flow of digital information soothes with its fam-iliar creep, the user's fix takes the edge off the experience of the world opened by the Great Hall. This latter experience is attenuated as access to the space of flows and the network of digitized information is dra-matically opened. The information onscreen is too compelling, the form of its mediation too seductive, for those who inhabit the on-demand world of screens to pass up. To be sure, the information delivered by the VUEguide succeeds in telling viewers far more or, at least, some-thing far different about the objects before them than they would know without access to the device and its content. The loss and gain in this exchange is difficult to measure. Provisionally, one might ask whether these are different modes of experiencing and knowing, and whether they are at odds. Augmented reality aspires to place a layer of data over the material world that leaves the latter intact and still sensible, but here, it seems, there is the possibility of an eclipse, in which the VUEguide and its network cast a shadow upon the world illuminated by the Great Hall and its things. As set out in a 2004 research report produced by Ubiquity Interactive – in which new media theorist Lev Manovich's notion of 'augmented space' is cited as inspiration – 'mobile devices and the mobile experience' concern not only 'ways of seeing' but also 'ways of being' (Ubiquity Interactive 2004, 8). This is what is at stake in the experience at the bottom of the ramp.

II

To use the VUEguide in the MOA is to confront the difference between experiencing the world as digital information accessible via wireless devices and networks and experiencing the world as revealed by a work of art in its place. In his essay 'The Origin of the Work of Art,' Martin Heidegger (1971, 15-86) argues that the essence of art is poetic. That is to say, the essence of art is the work it does to unconceal what *is*, the truth of beings and the world. Art, writes Heidegger (1971, 69–70), 'is the setting-into-work of truth ... the becoming and happening of truth ... the letting happen of the advent of the truth of what is.' Earlier in the essay, he puts it as follows: 'The art work opens up in its own way the Being of beings. This opening up, i.e., this deconcealing, i.e., the truth of beings, happens in the work. Art is truth setting itself to work' (Heidegger 1971, 38). Truth, for Heidegger (1971, 49), is *aletheia*, 'the unconcealedness of beings.' Unconcealedness happens only when 'an open place occurs,' when 'there is a clearing, a lighting ... That which is can only be, as a being, if it stands out within what is lighted in this clearing' (Heidegger 1971, 51). Clearing and lighting is accomplished in the poetic dimensions of art and thought. Truth happens in the work of art: 'One of these ways in which truth happens is the work-being of the work. Setting up a world and setting forth the earth, the work is the fighting of the battle in which the unconcealedness of beings as a whole, or truth, is won' (Heidegger 1971, 54). In art, an entity 'emerges into the unconcealedness of its being ... The nature of art would then be this: the truth of beings setting itself to work' (Heidegger 1971, 35). Heidegger famously refers to Van Gogh's painted depiction of peasant shoes. 'The art work,' he writes, 'lets us know what shoes are in truth' (Heidegger 1971, 35).

The poetic essence of a work of art is realized in *aletheia*, the unconcealing of the truth of beings and things. According to Heidegger (1971, 70), 'All art, as the letting happen of the advent of the truth of what is, is, as such, essentially poetry.' The essence of art is not realized in depiction, imitation, reproduction, representation or correspondence to apparent reality. Nor is it realized in information. It is realized in *poiesis*, in bringing-forth truth into unconcealment. In bringing-forth truth, by making a clearing for it and lighting it, art also presences the world. 'To be a work,' Heidegger (1971, 43) insists, 'means to set up a world.' And what is a world? 'World is never an object that stands before us and can be seen. World is the ever non-objective to which we are subject as long as the paths of birth and death, blessing and curse keep us transported into Being. Wherever those decisions of our history that relate to our very being are made, are taken up and abandoned by us, go unrecognized and are rediscovered by new inquiry, there the world worlds' (Heidegger, 1971, 43).

It is worth recalling that, long before the technologies of augmented reality, Heidegger diagnosed the challenge technological experience posed for what he described in his later writing as 'nearness.' We can only experience nearness, according to Heidegger, via an encounter with 'things.' In his essay entitled 'The Thing,' Heidegger (1971, 164) asks, 'What about nearness? How can we come to know its nature? Nearness, it seems, cannot be encountered directly. We succeed in reaching it rather by attending to what is near. Near to us are what we usually call things.' Placement, location, and nearness are materialized in our encounter with things, specifically, for Heidegger, things which 'thing,' 'stay,' or gather materially the fourfold of earth and sky, divinities and mortals. Absent a sustained, thoughtful encounter with such things, nearness collapses into its parody – the experience of distanceless distance and timeless time – producing what Heidegger (1966, 48) describes elsewhere as 'the illusion of a world that is no world.'

At the bottom of the ramp at the MOA, visitors cannot help but feel the nearness of things and be drawn by artistry into a world in which the fourfold of earth, sky, divinities, and mortals are gathered. This is what Heidegger (1977, 28) might call 'an original revealing.' Indeed, this spot may be one of very few in Canada where Heidegger's bizarre language actually becomes transparent and spontaneously meaningful. The contrast between the experience of the world presenced at the bottom of the ramp and that of the world of spatial augmented reality may be the difference between inhabiting a world of things and commanding a world of objects. Promotional material describes VUEguide as a tool designed to provide 'curatorial on demand multimedia' (Ubiquity Interactive 2004, 1). Is it possible that VUEguide takes the world of things opened at the bottom of the ramp and converts it into a standing-reserve of objects about which we can expect to be informed at our command?

III

This conversion is also suggested by a second recent attempt to augment an artistic experience in Vancouver by wireless technology. The 2006 Vancouver Sculpture Biennale saw twenty-two large-scale sculptures by major international artists installed outdoors at public sites throughout the city. Most were placed at locations along the scenic seaside walks that follow the shores of English Bay and Burrard Inlet, often visible at a distance from several vantage points, and approachable enough for climbing and touching. At Sunset Beach and Vanier Park, on opposite banks where False Creek exits into the ocean, stand Bernar Venet's *217.5 ARCS x 13* and *3 ARCS x 5*, two sets of massive (they weigh 5,500 and 2,700 kilograms respectively), rusted steel arcs welded together at precise angles and staggered intervals that, on this site, evoke the exposed ribcages of great whales haunting the harbour. At Devonian Park between Georgia Street and Coal Harbour, against a backdrop of Stanley Park and Cypress Mountain beyond, John Henry's *Jaguar*, a steel thicket of towering red sticks attempts the sky at twenty-five metres, even as gravity's hold on its 2,700 kilograms ensures the attempt can never succeed. At English Bay, in the heart of Vancouver's gay village, in a country that has recently legalized same-sex marriage, stands Dennis Oppenheim's *Engagement Rings*, two huge aluminum and steel engagement rings topped by illuminated glass solitaires, celebrating without apology the dignity and joy of the strolling couples whose place this most surely is. And on the south shore of Burrard Inlet, amid the gleaming glass and steel towers of the city's intoxicating wealth, a single bronze figure squats heavily, arms extended and sweeping. Ju Ming's *Tai-Chi Single Whip*, quietly defying the monuments to commerce, comfort, and technology that surround him, enacts an ancient practice that testifies to the many spiritual and ethnic diasporas that define this city and its futures.

In their awesome settings, these twenty-two resolutely material things open the world of this city in dramatic fashion. To use language deployed earlier, they bring the truth of the city forward into beauty and clear a space for judgment. Experiencing these works in their settings by walking and standing near them is something different than being informed about them, which is the aim of the free cell-phone tour augmenting the exhibition. Plaques beside each installation provide a local telephone number that connects to Ubiquity Interactive's Metrocode system. A code specific to the sculpture being viewed activates the audio commentary accompanying the work. The commentary takes the form of a casual conversation between a pair of erstwhile viewers, a man and a woman, played by two local improvisational actors. The dialogues – typically between two and three minutes in duration – are full of interesting information: details of the piece's fabrication, materials, history, and reception; the resumé and profile of the artist; possible artistic intentions and avenues of interpretation. The manner of presentation is defiantly populist, never didactic, and completely successful: it is witty, smart, engaging, and generous. In its most clever moments, the dialogue anticipates behaviour in which viewers enjoying such close proximity to these pieces are likely to engage. Just as an agile viewer, feeling rebellious, attempts to walk up the inside curve of Venet's *3 ARCS x 5*, she hears 'You know what I like to do? I like to walk up as far as I can before I fall backwards, then I walk backwards as far as I can before I fall forward' (and she wonders, just for a second, if she is being watched). Sitting on the grass watching two small children scale the joyous red ringlets of John Clement's *Kini's Playground*, the viewer plugged into Metrocode hears: 'This thing is a magnet for families and kids.' Following the dialogue, users are prompted with options to hear a list of 'Fast Facts' about the piece, leave a piece of commentary of their own (for possible posting on a website), or vote for which sculpture the city should retain from the exhibition. As Ubiquity Interactive co-founder Leora Kornfeld puts it: 'This is art for the people. The Vancouver Biennale is in the public domain, and Metrocode allows the public to use their cell phones to interact with the sculptures, get engaged and it makes art accessible to everyone' (Ubiquity Interactive 2006b).

What could possibly be wrong with that? Like VUEguide, the Metrocode Biennale audio tour is a brilliant technology, expertly executed with commendable intentions. Information is not all bad: it can tell you how much things weigh and suggest to you what they might mean. Still, one wonders whether the experience provided by Metrocode differs in kind from the experience of standing in the way cleared by the sculptures. An encounter with 5,500 hundred kilograms of rusted, perfectly geometrical steel arcs, placed mysteriously on a beach in the middle of a city, forcibly extracts people from the networks of digital information-on-demand, the space of flows and the experience of what Albert Borgmann (1984, 42-3) calls 'commodity.' The Metrocode cell-phone tour eases people right back into those networks, that space, and that experience. The losses and gains in this transaction are difficult to specify, but this much is suggested: one experience confronts us with a way of encountering the world that is radically different from the manner in which urban technological settings are now customarily inhabited; the other tends to confirm this latter as the normal way to be in the world.

IV

It is this difference – specifically, the viability of the abnormal condition of experiencing the world without recourse to information made available by digital networks – that is at stake in spatial augmented reality. Augmentation, it could be argued, is precisely the attempt to obliterate this abnormality, this difference, this other way of being in the world. The VUEGuide and Metrocode systems are quite basic attempts to produce the sort of 'augmented space' envisioned in contemporary accounts of augmented reality. As Manovich (2005) describes it, augmented space refers to 'overlaying layers of data over physical space ... augmenting this space with additional information.' Recognizing that there have always been ways of accomplishing this overlay of data on physical space (one thinks, for example, of signage), Manovich defines augmented space specifically in terms of augmentation by electronic or digital data, enumerating a list of technologies – cellular telephones, intelligent buildings and spaces, portable and embedded computing, pixilated screens, radio-frequency identification tags – that are symptomatic of a broad and dynamic range of steadily emerging applications of what is now often described as pervasive or 'ubiquitous' computing, digital media, and networks.

In their discussion of pervasive computing, Jerry Kang and Dana Cuff (2005, 95–9) identify three core elements of the augmentation of public space by digital information. These are: *ubiquity* (wherein digital networks are accessible anywhere, via a broad range of mediating devices and interfaces); *embeddedness* (whereby interface, computing, and network infrastructure are miniaturized and installed, virtually undetectably, in a wide array of objects in the material environment); and *animation* (wherein networked computing elements are capable of automated response to a broad range of physical stimuli, including biometric information). As these authors describe, implementation of pervasive computing will mean that digital networks 'will always be around – in the air and the walls – providing an ever-ready information template overlaid on the "real" world we navigate ... What we can expect, then, are networks of miniaturized, wirelessly interconnected, sensing, processing, and actuating computing elements kneaded into the physical world' (Kang and Cuff 2005, 94, 99).

Implementation of pervasive or ubiquitous computing systems oriented to realizing the various aspirations for spatial augmented reality is well underway, and now exceeds the integration of radio-frequency identification tags and tracking into the retail environment highlighted by Manovich (2005). Telephone companies, hardware manufacturers, Internet service providers, and search engineers are currently in frantic competition to roll out mobile social networking applications that enable users of portable, wireless, Global Positioning System–enabled communication appliances to locate friends, gather information on potential meeting places, and share directions (Hamilton 2007). The Tokyo Ubiquitous Network Project will install ten thousand infrared RFID transmitters throughout the Ginza neighbourhood to provide shoppers and tourists with wireless access to commercial information and navigation advice in four languages; a similar project in Tokyo linking cell phones to the Internet and GPS networks seeks to provide users with an electronic compass that includes step-by-step directions combined with detailed descriptive information and advertisements for over 700,000 locations throughout Japan (Williams 2006; Markoff and Fackler 2006). Researchers at the University of Guelph are presently at work compiling a database of small sections of the DNA sequences of every known species on the planet. In parallel, designs are underway to provide network access to this database from remote locations using a hand-held device capable of reading DNA samples scanned from organisms in the field (Jones 2007). The project, dubbed 'The Bar-Code of Life,' aims at a scenario in which any person, anywhere on the planet, will be able to identify any species and access information about it almost instantly. As the project's lead scientist, Paul Hebert, puts it: 'Any person equipped with a bar-coder can walk through the forest and identify the life around them' (quoted in Jones 2007, A7). Closer to home, academics might have had the pleasure of experiencing 'SpotMe,' a technology that enables conference-goers equipped with hand-held wireless devices to identify, locate, and send messages to fellow attendees with whom they would like to 'network' (Shockfish 2008).

Such applications are merely the tip of a very large, very sleek, iceberg. What augmentation of public, social, political space by pervasive computing and ubiquitous access to digital information via embedded networks will *mean* for the character of these spaces and our inhabitation of them is difficult to predict. It is certain that the affordances of these technologies will be deployed by corporate actors and state agencies for purposes of trade and commerce, the marketing and sale of entertainment and recreation, surveillance, and the enforcement of discipline and order. It is equally certain that artists, educators, and activists will seize upon these very same affordances in their own efforts to use these technologies to enrich our experience of the public sphere, and to craft new spaces of social and political encounter, in ways that encourage rather than discourage criticism, excellence, equality, and diversity. The technologies of spatial augmented reality thus bear a political ambivalence that is characteristic of technological systems more generally. Given the ongoing history of emergent media, it is safe to predict that every strategic deployment of these technologies for purposes of entrenching existing formations of socio-economic and political power will give rise to tactical appropriations aimed at contesting and subverting these very formations. It is even possible that, together, these 'secondary instrumentalizations' will shift the rationalization of spatial augmented reality in an altogether democratic direction (Feenberg 1999). It is entirely possible that the space of the Old City of Montreal might be augmented with data such that a tourist walking by Place Royale, scanning the screen on his cell phone for a map and review of a nearby restaurant, would also learn that, in 1734, a slave woman named Marie-Joseph Angélique was hanged there (Cooper 2006).

V

Still, while augmentation may be motivated by, or directed to, a variety of contingent purposes, there is nothing ambivalent about ubiquity. Ubiquitous, from the Latin *ubiquitas*, means everywhere and pervasively present. The Ubiquitarians were a sixteenth-century sect of Lutherans who believed Christ's body was present, everywhere, at all times. Ubiquitarianism was, and remains, a doctrine. For the Ubiquitarians of the sixteenth century, ubiquity meant the omnipresence and inescapability of God, and this was good. For the ubiquitarians of the twenty-first century, ubiquity means the omnipresence and inescapability of digital information, and this, too, is good. To the extent that spatial augmented reality is defined, at least doctrinally, by the goal of ubiquity, it is fair to say that it aims at an experience in which digital information, media, and networks are everywhere and cannot be escaped. It is in this sense that the achievement of spatial augmented reality might entail an eclipse of other, different ways of being in the world, regardless of the content, orientation, or application of the data it makes available. As Kang and Cuff (2005, 102) put it: 'Once implemented, opting out of pervasive computing will not be easy, and will eventually be seen as Luddite. After all, who among us regularly opts out of electricity, paved streets, security cameras, bar codes, web cookies or, in places like Los Angeles, even the automobile if we can afford one?' In the ultimate realization of spatial augmented reality, the standard rejoinder to the Luddite – 'if you don't like it, just turn it off' – is, by definition, unavailable. This is especially so when system ubiquity is combined with embeddedness throughout the material environment and automated animation by involuntary registration of biometric feedback (indeed, often by mere presence). It is at this point that spatial augmentation by digital information and networks becomes compulsory, the point at which it becomes, in a meaningful sense, reality.

To use a now disfavoured language, we might say that reality – in the sense of a compulsory framework of experience – comprises the *telos* of spatial augmentation by digital technologies – and that this *telos* is implied in all the diverse and contingent applications of the technology, including those that are democratic and those that are not, and those that fall short of completion in their actual deployment. It is with this in mind that one might ask: what sort of reality is it that is characterized by compulsory commerce with digital information, media, and networks? I would suggest that the character of the reality of ubiquitous, embedded, animated information and networks – the reality of spatial augmented reality – is indicated by the contrasting experiences of the space at the bottom of the ramp in the Great Hall of the UBC Museum of Anthropology, or the spaces opened by the sculptures of the Vancouver Sculpture Biennale, and that of the spaces opened by the VueGuide and Metrocode cell-phone tour. What can be detected in this contrast is the difference between inhabiting a world built upon compulsory enrolment in the flow of information and a world revealed in and by art.

One must proceed with caution in making such suggestions. Along with his reflection on art, things, and world, Heidegger had something to say about information. In a 1962 lecture on traditional and techno-logical language, Heidegger (1998, 139) contrasts information with language, or 'saying as showing and as the letting-appear of what is present and what is absent, of reality in the widest sense.' Information, by contrast, is an atrophied form of language proper to the regime of technology, 'the mere transmission, the reporting, of sig-nals' (Heidegger 1998, 141). Language – which Heidegger clearly identifies with the possibility of art – and information are radically distinct. 'That is why,' writes Heidegger (1998, 141), 'a poem does not, on principle, let itself be programmed.' This distinction might assist us in making sense of the world at the bottom of the ramp in the Great Hall and its difference from the world of information opened by the technologies of spatial augmented reality. For while the Great Hall and its things were surely made to communicate – the totem poles of the Gitxsan and Nisga'a peoples are media of communication even as they are works of art; the K'san doors are carved with a narrative ac-count the Skeena River peoples – it is arguable that the form of their communication is much closer to language than it is to information. This is why those who are not native to these languages (or to the lan-guage of Venet's sculpture), those who have never learned or been taught them, have to work very hard to understand whatever these things might be saying to them. Communication is difficult under the burden of language, but this burden can be lightened by information. This is precisely the promise of spatial augmented reality.

This promise, the promise of a life free from the burdens of art, lan-guage, and communication, can be evaluated on ethical, as well as technical and political, grounds. 'If,' writes Heidegger (1998, 141), 'one holds information to be the highest form of language because of its clarity, and the security and speed in the exchange of reports and assignments, then the result of this is also the corresponding concep-tion of the human's being and of human life.'[1] Heidegger goes on to quote directly the cyberneticist Nobert Weiner's *The Human Use of Human Beings*: 'To see the whole world and give commands to the whole world is almost the same thing as to be everywhere ... To live effectively means to live with adequate information.' This, we might say, is a premonition of the ubiquitarian creed of contemporary spatial augmented reality. However, it is an open question whether living ef-fectively is living well. This is the question raised by the recent aug-mentation of Vancouver.

NOTE

1 As one might expect, for Heidegger (1998, 141), the stakes here are equal
   parts ontological and ethical: '… as long as human being's relationship to
   those beings that surround and carry it, as well as to the being which it itself is, rests on the letting-appear, on the spoken and unspoken *saying*, the
   attack of the technological language on what is peculiar to language is at
   the same time the threat to the human being's ownmost essence.'

PART THREE

Locative Media

9 Labours of Location: Acting in the Pervasive Media Space

MINNA TARKKA

Much thought and action has been dedicated recently to the 'pervasive' media environment by artists, cultural producers, and theorists. Ubiquitous computing, broadband media, wireless and wearable applications, collaborative on-line platforms, and social software form a socio-technical assemblage transforming our spatial experience and opening up new potentialities both for regimes of power and for social inventiveness. This chapter visits some articulations of 'locative media' with the aim of drafting some critical context to this emerging artistic and technical practice. We are already familiar with the re-invention of spatiotemporal practices with mobile phones, from street-level user cultures, the spatial technologies of Geographic Information Systems (GIS), Global Positioning System (GPS), Radio Frequency Identification (RFID), to Closed Circuit Television systems (CCTV).

GIS databases, GPS positioning, RFID tracing, and CCTV networks add a totalizing grid and mesh of surveillance. Further, ubiquitous or pervasive computing involves the idea of 'invisible computers' embedded in objects and spaces, and 'smart' devices that can exchange information with each other over continuous networks and act together in a 'seamless' manner. The vision is to make technology calm and non-intrusive, to create 'environments saturated with computing and wireless communication, yet gracefully integrated with human users' (*IEEE pervasive computing* 2004).

The emerging landscape of *ubicomp* is thus an environment of translation where aspects of agency and 'awareness' of context are delegated from humans to machines, computational processes, and databases. Especially in cities, software is omnipresent as a kind of 'local intelligence,' infused into every fabric of urban life. Nigel Thrift and Shaun

French describe in detail the ways how software, through a series of performative 'writing acts,' contributes to an automatic production of space which conditions our existence by a continuous rewriting, standardization, and modulation of urban situations and rhythms (Thrift and French 2002).

However, instead of hegemonic or conspiracy theories of machines taking over, Thrift and French (2002) stress the contingent, distributed, ad hoc, and patched-up nature of this computing environment – a 'technological unconscious' rooted in the software cultures of programmers. This perspective prompts us to look closer at *practices*: it is through the mundane and minor, through everyday activities of programmers, designers, and developers that new forms of the social are being thought up and put into action.[1] With the focus on practices, account is taken of the tools and discourses of the work process, but also of the various kinds of invisible work and immaterial labour involved in the activity of production.

Of special interest here is the artistic and activist practice of the *ubicomp* environment recently subsumed under the banner of 'locative media.' It is crucially important that cultural producers intervene in this space whose parameters are set by the military and ICT industries: not only as 'early adopters' to develop cultural and social applications to new technologies,[2] but importantly, through their capacity to create new 'pervasive imaginaries' and to resist the totalizing tendencies and closures of *ubicomp* spaces. In addressing these labours of location, the key questions relate to how practices are positioned and negotiated within networks of culture, technology, and society. What are the tactics and strategies and how effective are they? How is the minor and mundane linked to grand narratives of progress in science and society? What kinds of potentialities, for thinking and acting, are performed into being?

Space, Place, Case (or Race)?

One camp is comprised of wild eyed zealots who are fervently convinced that we need to have freely available, machine readable, open licensed geodata, and will do anything to make that happen. The people in the other camp ... stare into your (wildly flashing) eyes,' their pupils dilate slightly and in a cracked bass exorcist monotone they say 'We have a very good relationship with the Ordnance Survey.' (University of Openness, Faculty of Cartography 2004)

'Locative media' is a loose common nominator for artists, developers, and activists who explore the possibilities of mobile, location-based, and other pervasive technologies. Their practice has presented a rich variety of projects, ranging from mobile imaging, sound, and performance to platforms for *moblogging* or *biomapping*, from exercises in *psychogeography* and *collaborative cartography* to experiments in *public authoring* and participatory *annotation of space*. The projects conceptualize and interweave the wireless frequencies of 'Hertzian space' with places and bodies on the move in the physical environment. Most often they use mobile devices or other components which through location and wireless technologies enable the production, transmission, and reception of media.

The writing that accompanies locative media projects involves utopian and dystopian reflections, playful and poetic manifestos, as well as programs for design and policy action (Russel 2002, Tuters and Smite 2004). As is typical of any media still in the making, there is a lot of 'weak rhetoric': a heterogeneous mixture of concepts, tools, and genres that are not yet aligned (Latour 1987). There is also the familiar romance with the 'new' in media; a passionate fumbling where a temporary loss of historical sense is combined with an archival impulse to search for antecedents and originators. In the case of locative media, the most often cited forefathers are Guy Debord and the Situationists, Gordon Matta-Clark, Michel de Certeau, Kevin Lynch, and Archigram. From this list, we can infer that locative media is about *urbanism* – perhaps the artistic counterpart to the emerging discipline of urban ICT studies proposed by Stephen Graham (2004a).

Not too many critical debates have taken place within the geographically dispersed locative community, connected by mailing lists and a chain of workshops and seminars.[3] Instead of a *problematization* of locative practices, the discussion has mostly been in the *problem-solving* mode, tackling with technicalities, proposing projects and collaborations, exchanging useful information. To work towards an analytic of the field – to decipher what problems locative media professes to solve – one perhaps has to locate some ruptures, breakdowns of consensus in the pioneering community.

Terms for such an initial debate stemmed from different understandings of space and place, a classical topic as such,[4] and revealed some *us-them* positionings, camps within the locative community. In the exchange, Giles Lane of Proboscis opposes locative media's inherent reliance on the abstract Cartesian idea of space, a 'desire to simply lock digital content to the most banal definition of place – i.e. the longitude and latitude coordinates that specify a location.' Instead of locations, he proposes to talk about *places*, seen as spaces of lived experience, social and cultural constructions (Lane 2004). Marc Tuters, a key spokesperson of the locative media network, responds by referring to network *topology*, in which the media should be distributed in a peer-to-peer mode and not stored on a central server, as is the case of Proboscis. This 'walled garden' approach is to be made obsolete by the semantic web, an esperanto for the Internet, and the creation of open source architectures, Tuters predicts (2004).[5]

So there seem to be two versions of ICT urbanism, the one more *ethnographic*, the other more *cartographic* in orientation, connected with different approaches to openness of structure and ownership of tools, but the discussion does not stop here. 'Locative is a case not a place,' Karlis Karlins from the Locative list reminds us; that is, the term was inspired by languages such as Latvian and Finnish with their several locative cases corresponding roughly to the preposition 'in,' 'at,' or 'by,' and indicating a final location of action or a time of the action.[6] In his posting, Karlins, who first coined the term, seems to be proposing a purification of locative media in an almost structuralist fashion.

The emphasis on linguistics is justified by the fact that much of locative development deals with semantics and formalisms needed for the description of space, the storage and retrieval of media, and the creation of algorithms. Here artists complement and contradict the evolving 'universal' standards for geography (OpenGIS) or navigation (W3C) mark-up languages by proposing more particular metadata schemes: semantics to describe mental maps, neighbourhoods, or psychogeography,[7] thus translating the social and cultural into machine-readable form, to the languages of software architectures.

Should we, then, approach locative media with geographic (*space*), social (*place*) or, linguistic (*case*) terms? All, I would say, and more. Interestingly, locative cases transform nouns by *inflicting* or inhabiting them; they have a performative force. For example, the six Finnish locative cases can, besides location and movement, also indicate time, causes and means, and even qualities, sensations, or relations of possession. This already allows a more *relational* understanding of location, one that is not treating locations simply as containers, 'in,' 'at,' or 'by' which 'content' can be placed. Locations also create asymmetries and 'localize' others, as Michel Callon and John Law (2004) point out from the relational approach of science and technology studies. 'The local is never local. A site is a place where something happens and actions unfold because it mobilises distant actants that are both absent and present' (6).

The question of *localization* brings us to some absences in locative discourse. Another debate thread was initiated on the Locative list by Coco Fusco's (2004) critique of the contemporary mapping-and-hacking enthusiasm, which 'evades categories of embodied difference such as race, gender and class, and in doing so prevents us from understanding how the historical development of those differences has shaped our contemporary worldview.' Locative media, as a new technology of localization, has been largely silent about issues of globalization, ethnicity, and gender, and about locative media's potentially colonizing effects on neighbourhoods (Appadurai 1996).

Promises of Participation

> Understanding the crucial relationships between people, places and things will increase our ability as designers and policy-leaders to suggest more open and people-centric uses of such technologies. We aim to create compelling scenarios and experiments demonstrating the benefits of authoring platforms that treat people as co-creative and not just consumers. (Lane 2004)

In working towards a critical contextualization of locative media, it may be useful to revisit the history of *site-specific* art. Miwon Kwon (2002) has pointed out how the label 'site-specific' became an uncritically accepted signifier of a critical and democratic art practice. By uncovering a genealogy of site-specific arts, she records the various uses – formal, functional, political – the concept has been accorded, while she surveys the movement from a more sculptural site-orientation to forms of institutional critique, community arts, and collaboration with local groups.

The focus on locations as locality, and the conjoined positioning of the artist as 'ethnographer,' are key elements also in locative media practice. Moreover, the practice is seen to be that of *collaborative* and *participatory* media. This turn, in new media, from 'interactive' to collaborative and participatory forms runs in parallel with a reconfiguration of social space, where the 'ubicomp' delegation of human agency to automated forms coincides with new regimes of governance and freedom. In the current model of 'good governance,' social responsibilities are increasingly delegated from the public sector and the government to communities and corporations. The weakening of the 'social' in society is supported by technologies for empowerment and self-management of communities and, in the end, for the responsibilizing of individuals (Rose 1999).

The ethos, often expressed as a morality, underpinning the emphasis on participatory media is that people should be liberated from being mere consumers and aided to become producers of their own content. However, the sharp opposition between producers and consumers has already dissolved in the contemporary cultural economy. The work of linking and chatting performed in social software environments, and the annotation of places in collaborative mapping and public authoring, are further examples of *immaterial labour* – the cultural, affective, and technical production that characterizes the contemporary 'social factory.' At one and the same time voluntarily given and unwaged, enjoyed and exploited, free cultural and technical labour is not exclusive to the so-called 'knowledge workers' but is a pervasive feature of society (Terranova 2000). The 'anytime/anywhere' of mobile communications is thus also the new quality of work, realized in collaboration with 'anybody/everybody.' Perhaps fittingly, the usual metaphor in geo-located messaging concepts is that of 'post-it' notes: the fetish of teamwork and brainstorming – of digital labour now made ubiquitous.

For Paolo Virno (2004), immaterial labour is best exemplified as *servile* labour, a work-without-end-product in which communication and cooperation are the main productive forces. This is exactly the situation in participatory media, where artists increasingly operate as *service providers*: their work becomes that of building platforms for user participation and collaboration, and of *maintaining* and *moderating* communicative situations. The continuous logic of collaborative value production also introduces challenging questions of accountability for the artist. If Kwon reminds us that communities are not only invented, but can also be exploited for the purposes of artistic and institutional career building, the creation of 'user' content in public authoring projects introduces new dilemmas of ownership. Should the virtuosity and 'linguistic performances' (Virno 2004) by users be considered as intellectual properties, gifts to the artist-provider, or voluntary services to 'community'?

There are several approaches to these questions entailing different theories of values and politics of collaboration and different models for public organization. A 'street' version of the Internet, locative media often subscribes to the discourse of early visionaries of the net and their contemporary legacy, the weblog theorists. Here the promise is of a participatory, open, and democratic media space, a space of creativity and freedom of expression. Once again, allegories for networked, collective intelligence and its alleged emergent result – an augmented, better version of participatory democracy – are fetched from the world of evolution. In a conjunction of biological and technological determinism, the flocking behaviour of animals is compared to self-organizing political movements and flash mobs. Attached to these metaphors is a selfish theory of action, 'power laws' to explain link economies, and micro-licences to enclose a 'creative' commons. There is a deep fascination with ant colonies displaying collective intelligence through emergent organization and pheromone trails (Ito 2004, Russel 2002, Tuters 2004).

The Uses of Visibility

I globally positioned the shadow of a cherry tree in blossom [N 56 56 648/E 024 06 646], chalking the coordinates on the floor inside the tree shadow, and writing the time from my GPS clock beside it. Then I wrote a haiku poem about it. (Gomes 2004)

Curiously, but perhaps not coincidentally, many locative art projects, especially those using GPS tracing, also evoke patterns of ant paths. Esther Polak's *Amsterdam Realtime* presents trajectories of GPS-deviced people moving about in the city, and Christian Nold's *Biomapping* adds an 'affective' dimension by visualizing galvanic skin responses – indicators of ease or stress – along the mapped trajectories. Despite their determinist appearance, both projects, however, aim to encourage the users' reflexivity towards their relationship with urban space by recording and exposing its patterns of use.

As already noted, a key feature of 'ubicomp' space is that its workings are largely invisible. The conditions of experience are being subtly changed from 'below' through algorithmic instructions, program runs, and database searches, and constrained by the immaterial spatialities of bandwidth and frequency. It is therefore understandable that *making things visible* is a desire shared by a variety of agents who seek to control, describe, develop, or resist the goings-on in this new space.

If surveillance classically was about the visibility of disciplined objects to the panoptic gaze, through technologies such as GPS, RFID, CCTV, and algorithms for face, gesture, and movement recognition, it has extended beyond the panoptic spaces of enclosure and become pervasive and *vectoral*. This is the society of control described by Deleuze, a system of variable controls which act to *modulate* behaviours, like a sieve whose mesh transmutes from point to point. Through the logic of code, 'individuals become "dividuals" and masses, samples, data, markets or "banks"' (Deleuze 1992, Rose 1999).

This technical administration of difference adds an important perspective to the issue of visibility: the question of information infrastructures. The 'invisible work' of categories, classifications, and data structures has very material consequences. When embedded in software, standards, archives, and infrastructures, they turn invisible and disappear into the uncontestable background of practices, from where they are not easily brought back for assessment or adjustment (Bowker and Star 1999). From this hidden background, they in turn operate on exclusion and inclusion, performing audits which again render some things visible and possible, keeping others out of sight and reach. The coming together of geo-spatial data with other types of metadata and statistics adds a totalizing dimension to this archive of behaviours, enabling pervasive techniques of social sorting, and increasingly acts towards closures and commodification of the public space (Graham 2004b). This is why the work of activist developers to free information infrastructures and open GIS databases is extremely valuable for 'us all data subjects' (Walsh 2004). Locative media connects agendas of public space with those of a digital public domain, and in this zone the urge to make things visible is connected to the urgency of access, of *making things public*.

From the design perspective, the need to make things visible relates to the immateriality of the phenomena – the 'beacons' that beam URLs, the geo-located messages suspended in 'mid-air,' the coverage of Wi-Fi nodes. The clear visibility of elements to be acted upon, and the provision of immediate feedback, are cornerstones of usability design, but how to make these happen off-screen, on the move? An example of wireless usability design is Matt Jones's *Warchalking*, a sign language to mark Wi-Fi hotspots on the street, which became a short-lived urban tech trend in 2002. Similarly, Pete Gomes's *Location, Location, Location* aims at creating future signage for the invisible via a conceptual architecture on the street. The use of chalk to mark streets is a low-tech version of a general visibility method, that of the *overlay*, where coexisting physical and virtual worlds are represented in relations of transparency, background, and foreground. The method stems from *mixed* or *augmented reality* applications which use data visualization or sonification techniques for layered representations of the virtual and the physical. Locative media artists employ the overlay to stage narratives, performances, and games in actual surroundings to produce embodied (mobile-aided) experiences which put virtual reality in your pocket, to be played back on the streets.[8]

In addition to these functional and expressive concerns of art and design, we also have the repertoire of counter-practices employed by many new media artists, who in the traditions of *détournement*, appropriation, and irony expose representational practices and politics by change of context, rearrangement of elements, or literalization of function. In fact, the many uses of visibility – functional, surveillant, descriptive, resistive – in the pervasive media space suggest the need to establish an interdisciplinary research field, a new visibility studies, to complement and revitalize the perceptually oriented agendas of visualization and usability research. Also, in this field of study, some critical questions of the effects and effectivities of visibility techniques, and of the links between the invisible and the affective, could be posed (Chalmers, MacColl, and Bell 2003).

The Tactical/Strategic Overlay

A walk / each day / in different shoes.
A walk / along a fold in a map.
A walk / without landmarks.
A walk / to the horizon beyond this page.
A walk / along an imagined line across your city. (Pope 2005)

What, then, are the visual practices involved in location-based arts? The 'locative' gaze conflates a god's eye view, the frozen military 'view from nowhere' of satellite vision and atomic clocks, with the situated, embodied 'pedestrian' perspective, the fleeting glance of the flâneur and the tourist, in search of consumable places and experiences. The two types of gazes coincide with Michel de Certeau's distinction between strategy and tactic. If *strategy* is about assuming a place to be isolated from its environment, acting on the objects and targets from a distance, *tactic* is a non-localized, temporal, and processual activity 'which insinuates itself into the other's place, fragmentarily, without taking it over in its entirety, without being able to keep it at a distance' (de Certeau 1984, 19). Besides overlays of physical and virtual worlds, 'locative' representations thus perform overlays of various power geometries. And of epistemic frameworks: Simon Pope discerns two modes of knowledge at play in locative media – a *sedentary*, static mode of maps and archives and an *ambulant*, mobile, in-between mode, that of walking art practices (Kwon 2002, Pope 2005).

Most often the overlay of these perspectives is represented through maps on which the geo-spatial hotspots are dotted and the users' mobile trajectories plotted. This type of conflation was presented in the *Cartographic Command Centre*, a collaboration of the Locative Media Lab, Project Atol, and others, in which maps of different scales, from satellite images to bio-mapped pedestrian paths and bicycled location video stream, were brought together in a stereoscopic 3D projection.

Military conversion, staging a 'centre of calculation' in an art context, may be a tactical act in itself, but in this project the overlay of geometries and perspectives did not add up to an oppositional message (Fusco 2004). Somehow the user experience was that of determinism; what was being staged was a spectacle and not a subversion of the all-powerful visibility techniques. The project exhibits a general problematic of the *tactical media* movement: its targeted micro-inventions often display an affinity with that which they seek to oppose (Lovink 2005). Maps remain strategic tools, technologies for governing at a distance, and their use for oppositional or creative purposes may just end in a reproduction of their spatiotemporal dynamics and structural logics, in a benign form of irony. It is thus all the more important to deconstruct existing mapping techniques, to question their ontologies and epistemologies, and to develop new formats for cartographic representation (Albert 2004, Rogers 2004, Sant 2004). Moving in this direction are projects in collaborative cartography such as *London Free Map*,[9] in which open-standard maps are redrawn from bottom-up by GPS-equipped walkers, cyclists, and skateboarders, projecting the tactical on the strategic.

A further dynamic, an interplay between determinism and chance, between *locating* and *stumbling*, seems to run parallel with the sedentary and ambulant modes. This is also where the inspiration of situationism, and the *dérive*, enters the game. Debord stressed that psycho-geographic drifts are not random, that they have a method of operating in the city and a political program of urban experience (Debord 1958), but locative practitioners have mostly used them as inspiration for 'disorienteering' expeditions. Thus Socialfiction's *.walk* takes 'method' to its extreme by presenting a human-executable algorithm for walking in, or stumbling on, the city. The algorithm (in its simplest form, '1st street left / 2nd street right / 2nd street left') should be able to produce a walk without navigational friction, but repeatedly produces more confusion than certainty. 'Technology will find uses for the street on its own' reads the project statement in a cyborg-ironic reversal of Gibson's slogan of street-level innovation.

Another locative genre combines the cartographic/sedentary and ambulant approaches in the creation of location-based public repositories. *Urban Tapestries* and *murmur*, among other projects of public authoring and urban annotation, invite users to participate by locating their own messages and stories in urban space, and audiences to follow the trails and threads signposted by these messages. In many ways, these projects are a further development of rhetoric, the art of public speaking and writing. They share a strong resemblance with the ancient art of memory, in which places (loci) such as streets or squares were memorized and used in the manner of wax tablet by orators who would place mental images on them in order to bring the speech topics back to mind during delivery. Whereas classical mnemonics was used in one-to-many situations – deliveries of political speeches or poems by orators – its locative many-to-many version promises an archive of lived experience, a community memory, or even a new type of 'commons.'

The participatory annotation of urban space fits well into de Certeau's description of tactical practice – for what else is annotation than a writing in the margins, a commentary which is never taking the space over in its entirety? Thus also: 'It has at its disposal no base where it can capitalize on its advantages, prepare its expansions' (de Certeau 1984, 19). The room offered for manoeuvre is not an empty container or wax tablet, but a space already configured by architecture, urban planning, and the telecommunication industry. Users operate within parameters created by these infrastructures and those of the platform-providing artists. In this context, it may be necessary to question also whether this work advances the specificity of sites, or the proliferation of *commonplaces* (Virno 2004)? The 'authenticity' that artists help communities to express is easily infused into programs of urban regeneration and place branding, as has been the case in previous local memory projects (Kwon 2002).

Location Economies, Location-Work

... they will be confronted with an image of their week, as well as the paths of the other participants. We register their reactions, ask questions, focus on landscape, politics, on their experiences and attitudes towards their surroundings, their perceptions of the potentials of the landscape, economic circumstances, myths about space, local songs, family relationships to the land. (Auzina 2003)

Urban locations, with their 'creative' demographics and 'authentic' experiences, are a prime source of value for contemporary capitalism, and since the great spectrum auctions at the turn of the twenty-first century, location-based services have been cast as a key 'value-added' in mobile telecommunications. Services for routing and tracking, fleet management, pervasive games, and user-profiled target markets figure strongly in the industrial imagination. The demise of the 'commons' gets a new turn in this 'dividualized' landscape of push-and-pull services. Besides the questions already raised concerning spectrum allocation and ownership, we now have to ask whether we will still have rights to our own location in space and time, or the trajectory we perform through movement? After all, other immaterial phenomena, such as oral traditions and genes, have already been brought under patent regimes. The colonization of new spaces, as before, takes place through translation, formalization, and mapping.

We can also already decipher new fetishisms forming around the production and consumption of place, and locative media participates in their production. Fetishes are made of immaterial social and spatial processes when their tropes, such as links and maps, are taken literally for the thing itself. Donna Haraway (1997) discusses genetic maps as 'ways of enclosing the commons of the body – of *corporealizing* – in specific ways, which, among other things, often write commodity fetishism into the program of biology' (148). Could we, in a similar vein, interrogate the 'localizing' effects of current locative practices which, like Haraway's gene maps, seem to defend 'the subject from the too-scary sight of the relentless material-semiotic articulations of ... reality' (Haraway 1998, 190)? Scanning through the emerging canon of 'locativity,' it is surprising to see how the very *context* (awareness of which is claimed as a key element of the practice) is often bracketed out in a reductive move from spaces to maps, places to dots, and sociality to links. The locative 'cartographers' shy away from the dirt and materiality of everyday life and prefer a resistance-at-a-distance.

If there is a certain degree of romanticism in these gestures of cartographic and psycho-geographic subversion, the same can be said about the locative 'ethnographers,' whose engagement with and empowerment of local communities run the risk of becoming functional reforms for governance-through-community or nostalgic evocations of authenticity (Kwon 2002). Meanwhile, anthropologists and ethnographers themselves have for some time problematized their spatial practices, especially with the discipline's foundational concepts of *field* and *site*. In this view, sites are not primarily spatially determined: the ethnographer's practice is multi-sited, a 'location-work' placing attention on the practitioners' social and cultural location and creating epistemological and political links with other locations. These situated, partial perspectives have also been applied to a feminist re-imagining of technology production in terms of *located accountability*, which stresses the careful negotiation among users, producers, and heterogeneous contexts (Gupta and Ferguson 1997, Marcus 1998, Suchman 2000).

Examples of such multi-sited practice can be found in the locative arts as well. In *Situations* and *Imaginary Journeys*, Heidi Tikka (2005) exposes the relative immobilities of families with small children. Through temporally arranged sequences of mobile images, her projects bring to view the rhythms and repetitions of families' spatial practices and explore ways to re-imagine private geographies in collaboration with distant actors. Esther Polak's and Ieva Auzina's *Milk* follows the routes of dairy production from rural Latvia to the Netherlands, tracing not only the GPS paths leading from the milk farms to the cheese gourmand, but also the changing networks of food production in the European Union. By employing simple documentary means of narrative and photography, these projects manage to connect mundane aspects of everyday life with the wider 'cosmopolitical' contexts of the global economy. In their minor and partial ways, they also describe hybrid economies of space, time, and location, negotiated across sites of the material, the social, and the technological.

This chapter has provided a brief overview of the cultural and technical practice of locative media. Partially, often tactically, with connections to the art world, academia, and industry, these developments mostly take place across trans-local networks, mailing lists, and workshops. Most of the work is immaterial and processual, and its results are visible so far only for a relatively small group of artists and developers. But it is symptomatic of a wide-reaching reconfiguration of social space and of a new sensibility towards the urban media space.

Locative media's passion for maps and their ontologies, networks, and topologies; the urban drifts to trace human geographies, the capture of media to record the murmur of the multitude – these are not only symptoms of a 'Red Dot Fever,'[10] of novel spectacles of 'you are here.' By building do-it-yourself transmitters, devices for mobile sensing/acting in a mesh of wireless nodes and networks; by inviting public participation in the production of media and infrastructures, this practice of 'next little things' also challenges the more monumental

forms of computational ubiquity (Lichty 2004). Locative media participates in the current problematization of place dominated by agendas of
ubiquitous computing, wireless telecommunications, and economies of
creative regions. Its speculative, empirical, and inventive practices contribute not only to a new study of ICT urbanism but also, perhaps, to
deciphering an affective politics of the urban.

NOTES

A previous version of this article appeared in *Species of Spaces*, ed. Giles Lane,
Diffusion eBook series 2005 (London: Proboscis, 2005).

1 This is the approach of the history of the present as proposed by Rose (1999).
2 Galloway (2004b) reminds us about *ubicomp* originator Marc Weiser's
   vision to frame the research field in cultural and social terms, an orientation
   which was later more or less effectively dropped from the agenda.
3 [Locative] http://base.x-i.net/mailman/listinfo/locative, [New-Media
   Curating] http://www.jiscmail.ac.uk/lists/new-media-curating.html. In
   2004 dozens of workshops on locative media were organized, and the issue
   was foregrounded in most international media art festivals.
4 For earlier discussions on the dynamics of space and place, see, for
   example, Lefebvre 1991, de Certeau 1984, and Augé 1995.
5 The exchanges between Lane and Tuters took place in various forums
   during early 2004, inaugurating the 'hot summer' of locative media.
   Us-versus-them positions – internal struggles over distinctions as indicated
   in the quote that opens this section – are common even within small fan and
   avant-garde subcultures. See Hills 2002.
6 Karlis Karlins, posting to the [Locative] list, 10 May 2004.
7 For proposed metadata schemes see, for example, Jo Walsh's *mudlondon*,
   http://space.frot.org/; NML, Neighbourhood Markup Language, by David
   Rokeby, http://proboscis.org.uk/prps/artists/rokeby/nml5.html; and
   Socialfiction's PML, Psychogeographic Markup Language, http://www
   .socialfiction.org/psychogeography/PML.html.
8 For example, Teri Rueb's *Itinerant*, http://www.turbulence.org/Works/
   itinerant/index.htm; *Uncle Roy All around You by Blast Theory*, http://
   www.uncleroyallaroundyou.co.uk; *Sonic City*, by Interactive Institute,
   www.tii.se/sonic-city/. See Galloway 2004a for a discussion of playful
   mobilities and Galloway 2004b for an extended discussion into
   embodied performances.

9 http://uo.space.frot.org/?LondonFreeMap. London is perhaps one of the
   most mapped cities in the world, especially as it comes to 'pedestrian versions.' The most famous cartographer is Phyllis Pearsall, who trod the
   streets to produce the *London A-Z* atlas. *Consume.net* has mapped the DIY
   Wi-Fi nodes; Proboscis weaves *Urban Tapestries* in Bloomsbury; while the
   London Free Map extends its streetnet from East End on.

10 'Red Dot Fever' is a skeptical response to the proliferation of urban
   annotation projects quoted by Albert (2004). After finalizing this article, a
   new debate formed on the discussion list of the Institute for Distributed
   Creativity. The thread on 'Interactive City: Irrelevant Mobile Entertainment' voiced a strong critique of the consumer spectacle related to many
   'playful' and performative works of locative media.

10 Spectrum Policy as Art: Interview with Julian Priest

INTERVIEWED BY BARBARA CROW

Julian Priest has been a critical player in the development and implementation of open wireless networks. In particular, Priest has been interested in the politics of spectrum and its implications for community and public use.

This interview describes Priest's policy intervention via an installation, *The Political Spectrum*, designed and orchestrated for Visual Spectrum, held in 2006 at the National Art Museum in Latvia. This piece provided visual representations of the electromagnetic spectrum. These visualizations often take the form of a spectrograph – a long thin strip marked with bands of frequencies. This image is formed in the physics laboratory by shining visible light through a prism or diffraction grating and shows bands of emission and absorption of light by different elements. The electromagnetic spectrum diagram extrapolates this form into the non-visible frequencies. Spectrum managers use a version of the spectrograph called a frequency allocation table to show the use of radio frequencies by a social group. The table simultaneously makes a visual claim to a natural, physical order and presents a map of 'The Spectrum' as a territory that can be divided and ruled. However, Priest's presentation of the visualization, and his invitation to visitors to participate in the commentary and design of spectrum, produced a social document of the names of all social groups that have won rights to emit and absorb communications over the course of the last century or so. To Priest's surprise, the spectrum table and white-board exercise revealed some of the changing political processes in Latvian politics and spectrum policy more generally.

This interview was conducted 17 and 22 November 2006 and 22 February 2007.

a) Please describe the spectrum project you installed in Latvia.

The Political Spectrum was exhibited as part of the WAVES exhibition in Arsenals Hall at the National Art Museum of Latvia. It was realized as a 6 metre by 1.5 metre white board attached to a long curved wall in the exhibition hall. At the opening of the exhibition, the white board was blank, with just a card describing some of the themes of the piece next to it.

A white board is a model of 'The Spectrum' where I am trying to suggest that 'The Spectrum' is less like a place to be divided up like other common resources such as land, but more like an infinitely rewriteable medium, a space of endless possibility, the space of all possible information transactions. It is a medium where you can both transmit and receive, and in the exhibition space the viewer's visual cortex is filled with the white board, so literally the white board represents the possibility of any potential image that can be viewed – the visible spectrum.

During the two days of the WAVES conference, which opened the exhibition, we drew up a chart of eleven long strips running the length of the board, one strip for each country in the local Baltic region. The countries chosen are members of a Baltic Economic Development Region of the European Union made up of countries bordering the Baltic Sea.

In each country and region, use of the electromagnetic spectrum and radio devices is strongly regulated by national governments, regional bodies like the European Union (EU), and international treaty organizations such as the International Telecommunications Union (ITU). This is traditionally done by assigning radio frequencies to each different use or user group within the geographical area of the regulatory body. For instance, when you tune a radio, you tune through different frequencies, getting a different station or user group for each one. This way of splitting up radio usage rights by frequency continues in a similar way for much of the electromagnetic spectrum through all the different uses such as television, microwave ovens, and even things like avalanche detection equipment. All these uses and their frequencies are held in documents called frequency allocation tables at each national, regional, and international administrative level.

This process of allocation has been going on for nearly a hundred years, and there are literally hundreds of thousands of entries in these tables. Each allocation in a table represents both a use and its user group, who have been through some political process to establish their right to use the public resource of radio. So frequency allocation tables are a kind of fossil record of the political successes of different radio user groups.

Rather than focus on 'The Spectrum' as a physic's object, I wanted to look at it as a social construction and to get some idea of the relative political success of different spectrum users in different countries in the Baltic region. I looked for ways to represent the data in the allocation tables without reference to radio frequency.

As part of the European Union's ongoing policy harmonization process, the European Radiocommunications Office (ERO) has developed an excellent spectrum allocation database, EFIS (ERO Frequency Information System). This produces on-line tables with standardized English terms for each European regulator for particular uses, making it easy to do queries and searches and comparisons.

Starting from this equation (well, slogan really), Frequency = Power, I did a text analysis of the frequency of these standardized terms in each of the Baltic countries, which gave me the number of times each user group had successfully been allocated a frequency band. It does not show how much bandwidth they have been allocated, but just the number of entries. This is a better measure of political power of each group because it shows how many times boundaries have been set or boundary disputes settled by each group.

Arranging this data from left to right on the white board, we drew the terms up on a logarithmic scale, so that a user group with few entries would appear on the far left and one with a lot of entries would appear on the far right. This lead to some interesting results – user groups like 'Detection of Avalanche Victims,' with little political power, might have a single entry, whereas 'Navigation Systems' or 'Military Systems' might have two hundred or more. The chart allows you to see relative influence of groups within a country, but also relative influence between countries. For instance, 'Detection of Avalanche Victims' appears once in Estonia – a country with no mountains – but over thirty times in Germany, with its mountainous alpine borders.

I also included the regional allocation data for the EU and the international data for the ITU region C[1] to locate the charts in their global contexts. The strip for Latvia was intentionally left blank, both as its inclusion in EFIS was then just under way after the recent accession to the EU of the Baltic states and also to leave space for the process-based part of the piece to unfold. This background data layer represents the current state of affairs, where the different political policy processes have got to.

The second stage of the piece explores how we create this particular structuring of our bit of the infinite space of informational possibilities through different administrative processes using metaphor and some public participation processes.

The white board was located in a special place – the National Gallery of Latvia – and this framing space represents the State. The curators who kindly allowed me the opportunity to display the piece are a bit like the government empowered to make decisions as to how the public interacts with the space.

As an artist, I took the role of civil servant or regulator, representing the current state of policy on the white board and mediating public write access to the medium through four different processes drawn from contemporary spectrum regulation.

In the middle section of the board, I drew two lines and delineated the area between them as unlicensed – anyone could draw anything they wanted in this area. This represents the policy approach of licence exemption used in 2.4 GHz for many user-installable devices such as Wi-Fi and Bluetooth.

In the blank Latvian section to the left of the board, I announced a beauty contest[2] used to allocate spectrum. I invited people to put suggestions for the use of this area in a suggestion book and attempted to form a committee from curators and artists to make the decision as to its use.

In the blank Latvian section to the right of the board, I announced a spectrum auction to take place late on the public open night. Spectrum auctions are a recent market-based management technique used most famously in the 3G auctions in the UK to attempt to get the market to set the best price for spectrum. This piece of the board would be auctioned to the highest bidder.

Lastly, I supplied some coloured pens which people could use anywhere on the board as long as they respected the lines of others. These represented modern technical approaches to spectrum management, such as smart radios or software-defined radio that actively seek out white space that no one else is using.

On the opening night, I invited the curators to pull up a section of the barrier tape and declare 'The Spectrum' open for use. I offered people pens and explained the regulatory options to them.

While the show was very busy, some people were slightly intimidated, perhaps by the presence of me as regulator, or perhaps by the idea of transgressing the gallery space taboo of touching the artworks, let alone adding to them. Nevertheless, gradually the unlicensed bit of the board began to fill up with a few bold souls taking the coloured pens and writing into the white space between the lines in other areas of the board, or adding to the suggestion book for the beauty contest.

As the evening went on, I realized that we had been drawing all day and had not eaten, so we went out to grab a bite to eat. As the piece seemed to be developing well and the regulations were being followed, we wrote up the rules for each section in both English and Latvian, announced the time of the beauty contest and auction, and left the box of pens open under the piece.

When we returned an hour later, we found a crowd around the piece, people of all ages writing messages, tags, signatures in all different colours in every available space on the board. The Latvian strip especially was filled with a dense thicket of writing and drawing. The exhibition visitors had understood the piece as a public space of expression, ignored the complicated regulatory arrangements, and made the piece their own.

Faced with the option of explaining the regulations to the crowd and clearing the non-compliant additions with a board rubber, I respected the regulator's principle of prior occupation and abandoned the auction and beauty contest.

What surprised me was the emotion I felt seeing the mass of people crowding around the piece and ignoring the carefully constructed regulatory processes – taking authorship. The whole point of the piece was to be a public space, but I was shocked that over two days of drawing I had come to have a sense of ownership over it as author and regulator. Over the next hours, I sat and watched how the piece developed and how people were making it theirs, and let go of my attachment to revel in the plethora of responses and uses that people made. (Check out the correct logarithmic positioning of 210 on the scale that someone added!)

I noticed that rather than a free-for-all, people largely respected both the existing data structure and each other's lines, normally choosing to work into available white space. The piece was exhibited and added to for a further month and is now a dense forest of drawing. It is currently hanging in the Department of Public Administration in the Economics Department of the University of Riga.

b) What made this project unique in your political work on spectrum?

Most of my work with radio spectrum has been through direct engagements in policy forums, writing consultation documents, and organizing meetings with regulators or public forums for discussions of spectrum politics. Initially with Consume and more recently with OpenSpectrumUK, we have had high quality interactions in spectrum policy debates and formation processes. Nevertheless, I personally always felt like a fish out of water and hopelessly under-resourced compared to the main industry and public players. Previous policy engagements also tended to feel oppositional or confrontational, trying to suggest directions for change in the face of entrenched positions and interests. While we have certainly had impacts in raising awareness within the debate, we have always worked in response to other initiatives on someone else's terms.

By exploring these issues as an art piece, I was able to step outside of that debate and explore the issues on my own terms without having to be bounded by the political process. While not my intention, this fed back well into the policy world.

After the show was over, I wanted to find a Latvian institutional home for the piece, as by then it was largely the work of Latvian people. With the kind help of Dr Aivars Kalnins, I was able to donate the piece to the Public Administration Department of the University of Latvia.

I also wanted to somehow reconnect the piece with the policy environment that it came from, and Dr Kalnins helped me to find some people within the Latvian regulator Education for Sustainable Development (ESD) for a reaction. In a telephone conversation with a representative of ESD, I asked for an opinion about the piece and the WAVES show, and was met with a defensive and cautious response that was very familiar from my policy engagements – being treated as an outsider. I nevertheless continued to tell the story of the piece, but when I got to my shock about the overrunning of the piece, by the public, the tone of the conversation changed. He commented, 'Now you know what it is like to be a regulator,' and we immediately entered a warm, frank, and fascinating discussion of spectrum regulation in Latvia and its future.

So the piece opened up a means where we were able to stand in each other's shoes, and that has given quite a different level of dialogue from what I have been used to in formal processes.

c) What was the impact of this work for you as an activist and in a policy context?

Later on in the process of reintroducing the piece to the policy world, I approached the ERO, who created the EFIS database I used, and had the opportunity to present the piece to them. There I also had a very productive conversation about the issues surrounding the piece, and they were very supportive of the work. This opened up some possible collaborations, and shortly afterwards they asked if they could exhibit a print of the piece at the 13th European Conference of Postal and Telecommunications Administrations (CEPT) conference in Berlin, which ERO hosts. This conference is the major meeting where all the telecoms and radio regulators in Europe meet to discuss upcoming policy issues, and to have the piece presented there to this audience by the ERO was a great privilege. I organized a 2 metre photograph of the piece to be shown, which now hangs in the foyer of the ERO offices in Copenhagen. So what started out as an experimental artistic exploration for an arts audience in fact was picked up inside the policy world and functioned as an awareness-raising piece and discussion point. Oddly enough I think it is the most successful policy intervention I have made.

d) What have you learned from this kind of intervention?

I really enjoyed the actual making of the piece and learned a lot from that. The participation and conversations with all the different people who helped to make it was fascinating – so many great perspectives and reactions, and I learned a lot about what my role was in the whole process of it, much more like a facilitator than an author.

I also learned that it is possible to bridge between art and policy, that you can use policy process as material in art pieces and that you can explore policy issues through art and that the results can be stronger in both domains as a result.

e) What can we learn about the changing political processes that have formed spectrum by exploring alternative visualizations?

Spectrum regulation is in a state of change. Radio technology is developing extremely rapidly, and the ways in which boundaries are set between radio users with fast switching, dynamic frequency selection, signal processing, higher frequencies, and smart antennas are opening up new ways of real-time organization of radio transmission, which creates more space to communicate in.

At the same time, the regulatory framework is in a slower but steady state of change, with old command-and-control structures being replaced by market mechanisms, and developing experimentation with technical and protocol-based approaches in licence-exempt spectrum and elsewhere. Simultaneously there is an explosion of radio devices and a great demand by users, as we have seen with wireless Internet. This is part of a great trend in the inversion in media from a broadcast model to a bottom-up or peer-to-peer model spurred by the Internet, making transmission a mass phenomenon.

With the huge variety of established uses of radio and vast numbers of 'legacy' devices in operation, it is neither probable, desirable, or even perhaps possible for radio spectrum to become an adaptive open spectrum – the existing structure is so firmly embedded in all our ways of doing things that change is likely to be generational, both in terms of people and technologies. The way this all plays out is being hotly contested by companies, governments, and engineers, but is not really visible to people until new generations of gadgets come to market.

In all the spectrum policy work I have been involved with, I have always run up against the problem that people cannot see spectrum. You literally do not see radio, but also the whole policy process is particularly obscure – hidden in device implementations and procedural government documentation, but most people are not aware of spectrum politics particularly.

When people do think of an image of spectrum, it is the physicist's rainbow strip, but this paints a misleading picture of spectrum as territory that limits our imagination of what is possible.

By making the political structure visible for people, and demonstrating people's desire to communicate in unrestricted ways visible to regulators, I hope to have developed a tangible image of the tangle of social and political processes that structure the physicality of our information exchange – The Political Spectrum.

This installation was shown at the 'Art + Communication' festival, organized by the RIXC in Riga, Latvia, as part of the WAVES exhibition, 25 August – 17 September 2006, Exhibition Hall ARSENALS, Latvian National Museum of Art, http://rixc.lv/06/en/press.html and http://rixc.lv/06/en/11.html.

Exhibition details: The Political Spectrum, Julian Priest, 2006; policy advisor, John Wilson; 5 metres x 1.5 metres dry erase marker on white board.

Acknowledgments: John Wilson, policy advisor; James Stevens, dataset drawing; Rasa Smite, curation; Raitis Smite, curation; Armin Medosch, curation;

Exhibition visitors in Riga, drawing; Adam Hyde, drawing; Aivars Kalnins, donation; Simon Pope, advice; RIXC, event hosting; European Radio Organisation, dataset; ESD.lv, policy feedback; Pia Bloch, CEPT conference; Garry Hill, software; Gina Kupferman, printing; Juergen Neuman, housing; and New Zealand Embassy in Berlin and The Hague, travel.

NOTES

1 The ITU Council is made up of forty-six members from five different geographic regions, and they are elected by the Conference with consideration of fair representation of the five regions: (A) the Americas; (B) Western Europe; (C) Eastern Europe; (D) Africa; and (E) Asia and Australasia.
2 There are two primary ways that spectrum is allocated, through 'auctions' and 'beauty contests.' Auctioned spectrum goes to the highest bidder, and beauty contests require that applicants demonstrate that they have the best capacity to manage spectrum.

11 Augmented Urbanism: Locative Media Experiences in the Digital City

KAJIN GOH, MICHAEL LONGFORD,
AND BARBARA CROW

But the cities of today and tomorrow only exist within the twisted relations between the physical community and its network counterpart. We should think of it as a soft landscape that is constantly being updated by its users.
– Akira Suzuki, *Do Android Crows Fly over the Skies of an Electronic Tokyo?* 9

Digital Cities is a collaborative research project involving designers, engineers, and urban and communications scholars located at Concordia University.[1] This project investigates traces of the 'digital city' and its networks, from multimedia districts to virtual environments and mobile devices. The role of technology in the city has been an important focus of cultural research. While earlier studies have concentrated on such urban infrastructure as electrification and transportation routes, new communications media have given rise to technological systems and networks that reorder the city. Through the workings of new communications media, the social and technological apparatus of cities is transformed, altering the terms of urban theory and representation.

We are dividing this chapter into two parts. First, we will discuss *Digital Cities'* most recent project, *Urban Archeology: Sampling the Park*,[2] which was installed at Place Émilie-Gamelin in March 2005 and explored the social history of a city square in Montreal using sound, image, and Global Positioning System (GPS) sensors. This project examined the ways in which memory can be inscribed in space, drawing on field recordings, oral history, and archival material to form a deeply layered mediascape. Next, we will discuss the *Sonic Arboretum*,[3] a prototype for a multi-user platform in development. This project explores some scenarios for extending social interaction by integrating the potential of ad hoc networks linked to a database incorporating input from environmental sensors and creating a shared authoring environment.

In different ways, these projects explore the potential for social and technological interventions to reframe the role of public spaces as creative sites where play, collaboration, and exchange can extend across spatiotemporal boundaries. For the first project, *Urban Archeology*, we started by researching sites within the city that offered the following: a public space with a high degree of social interaction that could be viewed as both a node and network at the same time. We were also interested in choosing a site open to use in a number of different ways that could challenge any of the assumptions we tend to make about what constitutes 'wireless' space.

After some deliberation, we settled on Parc Émilie-Gamelin, sometimes known as Square Berri, nested in a prominent intersection where many forms of mobility are represented in one of the densest areas of Montreal. We liked the way in which the city square invoked the notion of the public commons. At the same time, it is not typical of many public spaces in urban centres. It is located somewhere between a park and something much more intensely urban. In terms of transportation, the park intersects Berri Street and Ste Catherine Street East, two thruways that divide the city north/south and east/west. It also provides access to the Berri-UQAM Metro station and the Greyhound Bus Depot. As well as serving as a major transportation artery, this site is a hive of activity, with the University of Quebec in Montreal (UQAM), Quebec's largest university, to the west, flanked by communications industries in the form of both public and private institutions, with CBC Radio-Canada to the south, the new Bibliothèque Nationale to the north, and Videotron to the east. The park is bordered on all sides by a diverse number of businesses: fast-food restaurants, a magazine store, cafés, and a mall complex.

The park itself is actually zoned as a 'place' rather than a park, giving it a different legal jurisdiction as far as the city is concerned. This means there are no bylaws restricting entry after 11:00 p.m., and as a result there is a twenty-four-hour cycle of activity, including an overnight homeless population as well as a constant buzz of surveillance by the police. During the day, the population is extremely heterogeneous, ranging from students, local business people, and service-industry workers, to skateboarders and squeegee punks.

Panorama, Parc Émilie-Gamelin, Montreal. (Photo: K. Goh)

As a case study for mobility, the site is challenging. Through Île Sans Fil (ISF)[4] there is free Wi-Fi access in the park, though there are more cell phones than laptops in use. There is a steady stream of pedestrians, though there is not much by way of seating. As an example of civic planning, it is not very well designed; each section of the park is discretely compartmentalized in a way that feels intended to restrict flow. Yet somehow, the park users are constantly adapting the architecture to their own needs.

Over time, we began to see the park as emblematic of a number of competing interests among the city, capital, small businesses, the nearby residents and institutions, and the different populations inhabiting the park itself. Our interest was to represent this contested space – through a combination of archival study and field recordings – as a kind of urban archeology, which would help render the 'invisible' visible through the medium of wireless technologies.

To create the installation, we used a new authoring environment developed by Mobile Bristol (MB), a research group at the University of Bristol, and HP (Hewlett-Packard) Labs in the UK. The Mobile Bristol Toolkit provides a 'drop & drag' GUI (graphical user interface) for attaching media files such as sound, text, and image to GPS coordinates. The authored experience can then be downloaded to an iPAQ,[5] or hand-held computer, and played back in real time and space using headphones and a GPS receiver. Using the programming layer of the software, the sound experience can be augmented and choreographed in a number of ways. For example, using 'source points,' we can create graduated sound experiences by changing volume from the periphery of a sound region to its central source point. You can also attach more than one sound to a location and have it play back in a specific sequence depending on one's location in the park. These layered sound experiences added a degree of dimensionality, creating the impression of walking through an event unfolding in space. In addition to the sound, image files can also be attached and played back on the iPAQ in a browser window or using Flash.

We opted to work with the 'Information-Documentary' and began a process of research and study that would inform content. We started from an historical perspective by gathering material from the city archives. Later, we talked to various individuals about the past and present circumstances of the park. We discovered that the park was the site of a Catholic mission for almost 120 years, up to the early 1960s. We located a nun, Sister Therese, the head archivist for the Sisters of Providence, who in turn introduced us to Sister Yvette. She worked in the mission for over twenty years and told us about how it included a soup kitchen, a pharmacy, a hospital for the poor, and a home for impoverished elder women.

We found microfilm articles from old newspapers, reporting how the mission was bought over in 1962 to make way for an envisioned 'Plaza of the East.' This project eventually fell through, leaving the site in limbo for almost twenty years, and at this time it served as a parking lot. Then, with an infusion of public parks spending in the 1980s – a movement to 'revitalize' the city – the park was re-landscaped to its present configuration, and was renamed Parc Émilie-Gamelin, after the nun who established the mission.

We then tried to integrate the archival material with current content through a practice of sampling: recording in situ and later mixing and shaping material into contrasting varieties of sound/media experience. Since it is hard to impose linear-type structures on locative media, we opted for a collage of voices and sounds to incorporate both abstract and literal material: oral narratives, recent events, hidden histories, and the diegetic sounds of the park itself.[6] For example, the user crosses a sound tunnel that conforms to the underground metro route cutting below the park, sampled and mixed as a kind of musical collage, with accompanying footage of its mid-1960s construction. On this side, the user will meet Sister Yvette, pictured on the left, and Sister Therese, archivists of the Sisters of Providence, describing the old mission, the location of the buildings, and the experience of living and working there. On the southern side of the park,

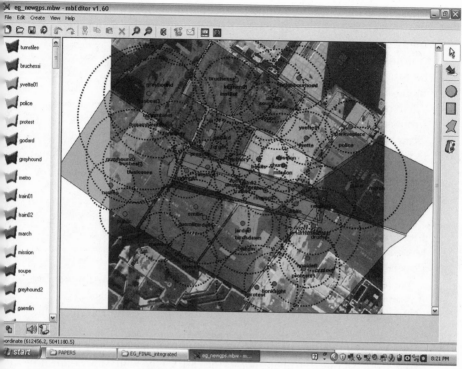

Screen shot: the Mobile Bristol Toolkit allows users to author a mediascape by assigning sound and image files to GPS coordinates. The file can then be downloaded for playback on an HP iPAC.

the user is immersed in the sounds of a recent student protest; as you walk this border section of the park, you follow the protest's evolution from a march down Ste Catherine to a mass sit-in at the intersection of Berri Street.

From Unidirectional Media to Server-Based Environments

One of the limitations of this project was that our content was in a fundamental sense 'canned' and unidirectional, relying on the traditional sender-receiver model in which the author creates and the audience consumes. In many ways, the experience of building *Urban Archeology*

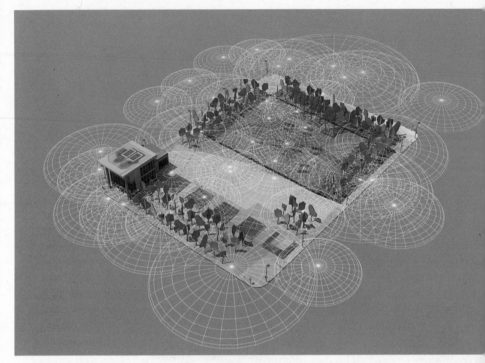

Urban Archeology: 3D map of hotspots in the park tagged with GPS coordinates. (Illustration: A. Morris)

was a prelude to a much more complex investigation. What we had done was build a prototype for content delivery that explored scenarios that served as a stepping-stone to an interactive platform that was server and sensor based.

The second project explored the expansion of locative media into server-based and streaming environments. The *Sonic Arboretum* is a proposal for a multi-user media-authoring environment in the guise of a gaming platform, fusing both virtual and physical space. It is in essence the idea of the 'play-space' embodied as social arena, designed to foster creative collaboration across spatiotemporal boundaries.

The word 'arboretum' refers to a space of cultivated trees and is re-

User in the park with GPS and iPAQ. (Photo: M. Longford)

flective of a central metaphor of our project having to do with ecology – the strategy of situating mobile communication activity within the larger framework of urban spaces as ecosystems, in which wireless networks would be more 'holistically' incorporated into the environment. This approach allows us to contextualize the flow of information within an expansive stream of other interactions: the flow of people, traffic, food, resources, energy, weather, and ideas. The ecological metaphor is a poetic methodology that has informed the shape of this project, employing the language of hybridity and germination, as well as ideas such as socio-cultural cross-pollination.

The *Sonic Arboretum* is a perpetual musical 'jam' feeding on sensor- and player-generated input, taking place in both the physical world of the park and the virtual world of the Internet. Players will be able to record and manipulate sound in situ, interacting with both previous and current players, while the environment itself, in the form of sensor-data streamed through a database, can be mixed and sampled, cut and spliced, and regenerated as a collective, open-ended score to be experienced both on-line and on-site.

The *Sonic Arboretum* is composed of two main elements: the Stage and the Score. The Stage comprises the park environment itself, and will function as an auditory template upon which users will interact and build their compositions. The Stage will be a generative sound landscape based on streaming environmental data collected by a network of sensors recording light and sound decibel levels, traffic and the proximity of pedestrians, temperature and humidity. This data will then be translated into variable sound values that will modulate sound files input to a database by the players.

This phase of the project is partly under way. We began last year to prototype a system of sensors capable of measuring basic environmental stimuli to be transmitted over a Wi-Fi network to the *Digital Cities* database on a central server and displayed on a website. The protective casings in which the sensors will be located are modelled as an extruded version of the Mobile Digital Commons Network (MDCN) logo. These 'bird' sensors, located at ISF locations, effectively turn each hotspot into a data-rich node in the network.

The Stage sets the basic parameters for the further layering of interactions known as the Score. The Score is essentially user-generated input, a constantly modulating, potentially endless musical improvisation that will also include static and moving visual material. The Score is based on the premise of the Open Artwork, or Art as Event, incorporating feedback loops and constant interchanges of material that can be both individually and cooperatively shaped over time. Against the backdrop of the Stage, which is partly artificial-intelligence driven, the Score embraces non-static principles of participatory creative endeavour.

Visualizing the Project

On the next page is a hypothetical model we have been developing over the year, a 3D rendition built by our CGI artist Antoine Morris. The eventual use of this model will be to generate a real-time, 3D environment through which on-line users can interact with players in the park.

In this environment, player presence is represented as an aura or energy field. In this example, physical, real-time players are tracked and represented according to their GPS coordinates. One of the things

Visualization – 3D rendering of the *Sonic Arboretum*. (Illustration: A. Morris, K. Goh)

that happens with GPS sensors is what is called 'drift,' meaning that on occasion the player loses his or her 'fix' to the various satellites triangulating their positions on the physical terrain. What the screen shows is a simulation of what happens when drift occurs, when the player's position suddenly shifts 20 to 30 feet off any of the cardinal points.

The somewhat phantom-like presence of GPS-based, real-time players is distinguished from remote on-line players. The blue figures represent on-line players who have logged into the virtual park, composed of randomly generating and regenerating bits whose movement is influenced by the navigational use of the keyboard.

Players log in either in physical or virtual space. The GUI (graphical user interface) allows them to generate lists of other players present at that particular time, as well as traces of recent players and their trajectories of files. Players connect by clicking directly on other players and dialoguing through an exploded menu. Interpersonal interaction allows options for text, voice, or video chat and access to personal sound, image, and video files. Player exchange is expressed through a 'docking' of windows; multi-user interactions produce multiple interlocking windows.

We have taken up an ecological metaphor and tried to apply it as a kind of behaviour determined by network activity. For example, we are interested in weather cycles – rain, sun position – and its influence on the underlying flora – seeding, growth, reproduction, and pollination. Heightened network activity will produce precipitation and clouds; areas of the park with especially dense network traffic will generate seeding, leading to the sporing of plant life.

Digital Arboretum: Game Platform

The plants – virtual trees springing up between the real trees – are actually inverted rhizomes serving as file-access points. The greater the differentiation of files, from points of origin to type of expression, the greater the level of represented hybridity.

As in an actual ecosystem, files are subject to a process of decay. In this environment, the database or archive is equipped with a kind of in-built mortality mechanism; without constant watering, the plants will wither and possibly die, though death is not figured as voided absence but rather as transmutation. The generation of files in the network is not static, but prone to continual mutation. Plant clusters will grow, wither, die, and spore new clusters.

The *Arboretum* is an attempt to extend the role of the database from the storage-and-retrieval archetype. Rather than a repository of 'information,' the anticipated database transforms the archive from an historical 'trace' into a live process. Operating partly as a communal palette for public use, the archive here becomes a tool for generating collective experience and ad hoc communities.

Conclusion

As both an electronic and an existing real space, the *Sonic Arboretum* reflects a desire to restore physicality to information space; rather than an aggregate of precipitate bits, the energy borne by the networks becomes embodied in the interplay between the remote and the local, the real and the virtual, played out in day-to-day life. It is a variation of the

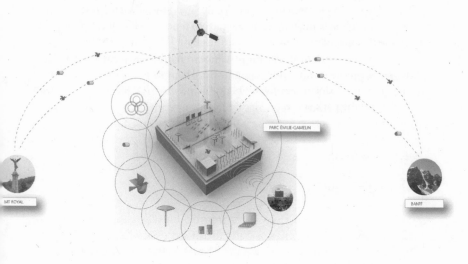

Wi-Fi/GPS network infrastructure for the *Sonic Arboretum*. (Illustration: K. Goh)

gaming experience in which there are no levels, tasks, goals, or directives. Focusing instead on play and random acts of creative exploration, it appendages social gestures into the fabric of its operation. Our project is an effort to realize what might be termed the 'social imaginary'; not an utopian model of possible interactions, but interactions exploring creative as well as social potentialities in the here and now.

NOTES

1 For details about *Digital Cities*, see http://www.digitalcitiesproject.net/.
2 The following individuals participated in the creation and production of *Urban Archeology*: Amitava Biswas, Barbara Crow, Kajin Goh, Ile Sans Fil, Jennifer Gabrys, Michael Longford, Bita Mahdaviani, Antoine Morris, and William Straw.
3 The following individuals participated in the creation and production of *Sonic Arboretum*: Barbara Crow, Michael Longford, Kajin Goh, and Antoine Morris.

4 ISF is a non-profit organization in Montreal that has installed over 150 'free' Wi-Fi hotspots, http://www.ilesansfil.org/tiki-index.php.
5 The iPAQ is a hand-held mobile device designed by Hewlett-Packard, http://www.hewlettpackard.com/country/us/en/prodserv/handheld.html.
6 While this installation shares similarities with Janet Cardiff's work, a sound piece in an urban space, Cardiff privileges a fictional experience of place, while our piece is interested in tagging media files to a specific location. We relied more on documentary strategies than Fine Art ones to convey meaning about the park space.

PART FOUR

Wireless Connections

12 The Wireless Commons Manifesto

HTTP://WWW.WIRELESSCOMMONS.ORG

In 2001 the Wireless Commons Manifesto was signed by key individuals who were experimenting with, creating, and implementing wireless strategies such as Wi-Fi networks to resist, and provide alternatives to, the telecommunication industries' domination of infrastructure. The original signatories included: Adam Shand (Personal Telco), Bruce Potter (CAWNet), Paul Holman (Shmoo Group), Cory Doctorow (EFF), Ben Laurie (Apache-SSL), David P. Reed (Open Spectrum), Schuyler Erle (NoCat), Matthew Asham (BC Wireless), Lawrence Lessig (Creative Commons), Jon Lebkowsky (EFF-Austin), James Stevens (Consume), Steven Byrnes (Houston Wireless), Richard MacKinnon (Rocksteady), Duane Groth (Sydney Wireless).

In the tradition of manifestos of revolutionary movements, from the women's movement to those of anti-globalization, the Wireless Commons Manifesto was an attempt to quickly, astutely, and politically lay claim to the possibilities of free, open, and accessible wireless networks in unlicensed spectrum.

We have formed the Wireless Commons because a global wireless network is within our grasp. We will work to define and achieve a wireless commons built using open spectrum, and able to connect people everywhere. We believe there is value to an independent and global network which is open to the public. We will break down commercial, technical, social and political barriers to the commons. The wireless commons bridges one of the few remaining gaps in universal communication without interference from middlemen and meddlers. Humanity is on the verge of a turning point because the Internet has transformed the way humans relate with one another. All communication can be traced to a human relationship, whether it's lovers exchanging instant messages or teenagers sharing music. The Internet has given us the ability to communicate faster and more cheaply than ever before in history.

The Internet's value increases exponentially with the number of people who are able to participate. In today's world, communication can take place without the use of antiquated telecommunications networks. The organizations that control these networks are limping anachronisms that are constrained by the expense and physical necessity of using wires to build their networks. Because of this, they cannot serve the great mass of people who stand to benefit from a wireless commons. Their interests diverge from ours, and their control over the network strangles our ability to communicate.

Low-cost wireless networking equipment which can operate in unlicensed bands of the spectrum has started another revolution. Suddenly, ordinary people have the means to create a network independent of any physical constraint except distance. Wireless can travel through walls, across property boundaries and through a community. Many communities have formed worldwide to help organize these networks. They are forming the basis for the removal of the traditional telecommunication networks as an intermediary in human communication.

The challenge facing community networks is the one limiting factor of wireless communication: distance. The relationships that can be formed across a community wireless network are limited by their physical reach. Typically these networks are growing to the size of a city, and growth beyond that point requires coordination and a strategic vision for community wireless networks as a whole. Without this coordination, it is hard to see how the worldwide community of wireless networking groups will ever merge their systems and create a true alternative to existing telecommunication networks.

There are many barriers to the creation of a global network. So far, the focus has been on identifying the technical barriers and developing methods to overcome them. But technical problems are the least of our worries, the business, political and social issues are the real challenges facing community networks. Hardware and software vendors need to understand the business rationale for implementing our technical solutions. Politicians need to understand our requirements for universal access to open spectrum. The public needs to understand that the network exists and how to get access. Unless these problems are identified and addressed, the community wireless movement will never have influence beyond a local level.

Most importantly, the network needs to be accessible to all and provisioned by everyone who can provide. By adding enough providers to the network, we can bridge the physical gaps imposed by the range of our equipment. The network is a finite resource which is owned and used by the public, and as such it needs to be nurtured by the public. This, by its very nature, is a commons.

Becoming a part of the commons means being more than a consumer. By signing your name below, you become an active participant in a network that is far more than the sum of its users. You will strive to solve the social, political and technical challenges we face. You will provide the resources your community consumes by cooperating with total strangers to build the network that we all dream of.

Adam Shand, Personal Telco, 2001

You might also like to sign the manifesto yourself or read our proposed definition of a community wireless network.
http://www.wirelesscommons.org/english/signatories

For an archived copy of the manifesto, please visit the following URL:
http://www.wirelesscommons.org/english/manifesto?redirect=1

13 Community Wi-Fi, Resistance, and *Making* Infrastructure Visible

ALISON POWELL

Introduction

As other chapters in this volume point out, the Canadian urban land-scape is increasingly multi-layered: the geographical space of the city is overlaid not only with the mental maps that urbanites produce as they move through it, but by other kinds of maps as well: the dense, pulsing maps of radio signals broadcast, transmitted, and received by the wireless devices that more and more Canadians carry. These layers of signals are invisible, at least to the naked eye, and few people take notice of them, except perhaps to complain when the service they provide is no longer available. Like that other technological marvel, electricity, wireless communications signals are quickly fading into the woodwork of the urban imagination. However, as they disappear, groups of activists and artists are trying to make them visible, to bring public attention to their provenance, and to suggest ways to connect the topography of wireless signals to more intimate local geographies.

By experimenting with, repurposing, and sharing Wi-Fi signals, these groups, called Community Wireless Networks (or CWNs),[1] attempt to make visible the invisible layers of communication infrastructure. This making visible is intended to draw attention to both the increasing con-solidation of Wi-Fi ownership, but also to the technology's potential to reinforce local community, culture, and practice. CWN members im-agine their networks as ways of 'hacking the city' – of reconfiguring relationships between individuals, communities, and communication networks in order to reaffirm the importance of local culture amidst global flows of communication. Therefore, CWNs are decidedly local, and employ different technical and social tactics in an effort to make

Wi-Fi infrastructure locally visible before its disappearance into the infrastructural invisibility of electricity and telephone service.

This chapter begins with a short history of community-based Wi-Fi endeavours, focusing on two waves of community wireless network development: a first wave concentrating on technical developments, and a second wave that includes wider social goals. Montreal's Île Sans Fil (ISF) – 'wireless island' – is introduced, along with its strategy of reinforcing practices and cultures of particular places by augmenting them with customized web pages and social networking tools that appear when an individual accesses the Wi-Fi network. A more thorough discussion of these tactics follows, explaining how ISF's aim of connecting the use of Wi-Fi to the culture of particular urban spaces and places attempts to perform a conscious integration of local cultural life and communication networks.

However, despite the success of CWNs – especially ISF – in making relationships between humans and networks visible through technical and social actions, their hacker's goals continue to be integrated into the political-economic realities of life under capitalism. Thus, while making Wi-Fi visible in public locations may create an opportunity to reorient its use, and thus the way it is collectively imagined, such efforts do little to undermine the material and economic control of the increasingly invisible communications infrastructure by telecommunications companies. However, this may be changing as the grassroots tactics of CWNs influence the development of hybrid forms of ownership and control of 'public interest' communication networks (see Clement and Potter 2007). While these shifts in ownership or control of communications infrastructure are important, the cultural legacy of CWN projects is less often considered. To address this, the second part of this chapter uses the ISF case to reflect on the future role of CWN projects in a context of invisible, ubiquitous Internet access. How can they continue to make communications infrastructure visible and part of a collective imaginary? Are these goals sufficient as forms of resistance or restructuring of the ownership of this infrastructure?

The History of Wi-Fi and 'Bottom-Up' Innovation

In 1999, Apple Computers introduced its AirPort built-in wireless Internet card and router, which received Internet connectivity via radio signals. This technology, built upon the interoperable 802.11b wireless standard, was the first commercial Wireless Fidelity (Wi-Fi) equipment.

Laptops with AirPort receivers sold wonderfully, permitting Internet users to unplug their Internet cables at home or in the office. Soon other companies offered Wi-Fi routers and receivers, all based on the same standard. Initially, these technologies did not offer very high-speed Internet access, and only extended about 100 m from the routing stations. Yet Wi-Fi was extremely flexible and, unlike other kinds of communication technologies, did not require creating a centralized network. Furthermore, any equipment would easily work with any other equipment, making the technology very accessible. These factors, combined with the high cost (at least in Europe and the United States) of commercial broadband services at the end of the 1990s, made Wi-Fi an attractive object for community groups who wanted to lower the cost of Internet access by sharing high-speed lines between friends and neighbours, and for hobbyists who wanted to experiment with new technological forms. As Auray, Charbit, and Fernandez (2003) point out, the technical potential of Wi-Fi seemed more suited to development of decentralized, alternative communications structures than to centralized ones. As the first Wi-Fi waves extended across the city, they bore the imprint of community groups and amateur technicians who, before commercial operators had a chance to define a 'Wi-Fi market,' created the systems and structures that helped to define what Wi-Fi signals mean for urban areas. The short history of the activities of these groups helps to situate the role of community actors, not just in appropriating Wi-Fi, but in defining it through their tactics.

Part I: Local Tactics, Global Networks

Since the early 2000s, community groups working with wireless Internet technologies have sprung up in nearly every major North American and European city. These groups, many composed of impassioned 'techy' volunteers, like the 'radio boys' of a previous wireless generation (Douglas 1987), took their inspiration from the flexibility and interoperability of Wi-Fi devices. Off-the-shelf Wi-Fi routers and equipment become raw materials for open-source software enthusiasts to program and redevelop into tools for sharing bandwidth, authenticating public Wi-Fi users, or sharing and presenting multimedia content. For example, many commercial Wi-Fi routers are programmed in Linux, the original open-source operating system language, leaving them open for modification by appropriately experienced community group members. Hacking, rebuilding, and remaking are the watchwords of this type of activity, and to those of us outside the CWN movement, they often seem more important in and of themselves than are their results. CWNs are innovators in local areas, where, in keeping with the open-source ideology of 'information should be free,' they create ways to share Internet connections and to provide Wi-Fi Internet free of charge to as many people as possible. In many cities, as Sandvig (2004) suggests, this leads to a proliferation of public Wi-Fi locations where Wi-Fi users do not pay for their connection. This creates a discursive connection between 'public Wi-Fi' and 'free Wi-Fi' that Sandvig argues reduces the willingness of people to pay for public Wi-Fi. On the other hand, it may also act as a way of framing Wi-Fi, and the Internet, as a public good – a good created and managed by a community group.

However, all CWNs do not concentrate on providing Wi-Fi in public places: the objectives and missions of these organizations vary, from providing a space for discussion of new technological developments by enthusiasts to creating a mesh network of Wi-Fi nodes that form an alternative 'intranet' network, not necessarily connected to the Internet. Since 2000, these primarily technical goals have defined CWNs, whose technically adept members have defined their activities in technical terms. However, the appearance of CWNs in Europe and North America has had as much to do with the state of telecommunication markets in each of these locations, as it has with the enthusiasm of volunteers, and as markets mature locally and worldwide, CWNs adjust their tactics. The following introduction of CWNs sketches out a quasi-historical framework in which Canadian CWNs like Montreal's Île Sans Fil who define public Wi-Fi as a means to develop and share cultural content, are framed as 'second-generation' networks, who have introduced socially oriented mandates along with the technically oriented ones of their American and European colleagues (see Cho 2006). This history indicates the extent to which social goals, always present in CWN projects, begin to be articulated in second-generation groups as primary motivations, even to the extent of shaping technical configurations.

First-Wave Community Wireless Networks: Consume

In 2001, London's *Consume* was the best-known CWN. Sandvig (2004) featured the group in his initial assessment of community networking, pointing out how its members attempted to create an ad-hoc network that existed in parallel to the Internet, as well as opening shared points of access. The idea was to open the Internet to 'everyone' while simultaneously creating a non-hierarchical, meshed structure that balanced out the hierarchies of Internet service operators and clients. At the time, the technical protocols developed by *Consume*, which allowed the network to be automatically configured based on the number of active nodes, were cutting-edge – they allowed the network to expand rapidly without human intervention. This technical accomplishment was perceived as being a way for Wi-Fi to accomplish what the wired Internet could not: democratize communications. *Consume* was also one of the first CWNs to participate in the Pico-peering agreement that codified the sharing of available bandwidth between European CWNs.[2] These primarily technical aims were shared by other early CWNs, some of whom wanted to use Wi-Fi uniquely as a way to create local communities, without providing Internet access. But whether oriented towards facilitating local information exchange or democratizing access to the Internet, these projects concentrated mainly on technical development. As Auray, Charbit, and Fernandez (2003) point out, *Consume*, and many other European CWNs, mainly concentrated on 'testing the limits of 802.11b equipment.' While some communities developed software to manage the use of their network, most concentrated on regulating network traffic and developing routing protocols that fairly distributed network resources – the paid broadband Internet connections of CWN members. These technical goals were supported by social relationships built between members in local CWNs and within CWN groups as a whole, but explicit social goals were not part of the first wave of CWN development.

Second-Wave Community Wireless Networks: Île Sans Fil

In North America, the first and best-known CWN was the Seattle Wireless Network, which hoped to cover all of Seattle with an alternative wireless mesh – a goal that resonates with that of *Consume*. The idea of an alternative mesh spurred the developoment of well-known American CWNs, not only in Seattle but in New York City, where NYCWireless famously covered Bryant Park with wireless signals, and in the university town of Champaign-Urbana, Illinois, where CWN pioneers created hardware and software that is used in both North American urban settings and in developing countries. In Canada, the National Broadband strategy has meant that urban CWNs have not adopted the same cost-reduction strategies as their United States cousins. Instead, Canadian CWNs have focused on technical experimentation that compensates for unique geographical conditions, and, most uniquely, on an integration of Wi-Fi into the cultural life of a city. Montreal's Île Sans Fil (ISF) pioneered this approach. A completely volunteer-run initiative, ISF has become the dominant provider of public wireless Internet access points, while growing as part of Montreal's community-based media scene. As the ISF website states, it is

> a non-profit organization dedicated to the development of a free communication infrastructure to strengthen local communities in the greater Montréal region. Île Sans Fil is both a technical development project and a grass roots community group, involving professionals and students from diverse fields including engineering, network management, software and hardware development, art curation, graphic design, sales, and marketing. The vision of the group is to use new technology, in particular wireless technology, to empower individuals and to foster a sense of community. (Île Sans Fil 2003)

As such, its goals attempt to transcend the activities of the group itself and instead begin to nourish local communities through the development of Wi-Fi hotspots in public places. ISF's unique focus on public space distinguishes it from other CWNs like *Consume*, whose primary goals focused on extensions of technical development. ISF aims to augment 'third spaces' away from work and home by providing Wi-Fi not merely as a service, but as a part of the experience of being in a certain place. As such, their approach makes visible the potential of Wi-Fi by anchoring it in the culture of a particular space, place, or location.

ISF members engage in two kinds of practices as ways of developing their vision of Wi-Fi. Like other CWNs, they establish Wi-Fi access points (or 'hotspots') in public locations across Montreal. ISF concentrates on locations that are open and accessible to the public, so that their hotspots are found in parks, cafés, bars, restaurants, artist and community centres, and the public areas of some hospitals and academic institutions. At each of these locations, the hosting institution signs a 'social contract' indicating that they will not charge any of their users for the service. The host also agrees to pay for equipment that ISF provides, at wholesale cost. This social contract codifies the relationship among the host, ISF, and the end users as 'social' rather than commercial, and attempts to undermine the way in which Wi-Fi becomes a commodity for consumption in public places. However, this same social contract depends upon host institutions paying for Internet service and being willing to offer it to clients without remuneration. As such, it is not a complete departure from commercial relationships, and Sandvig (2004) argues that it merely shifts the payment further back in the economic chain (for example, making end users unwilling to pay for the service and forcing businesses to absorb it into their overhead). To be sure, partners in the ISF project must pay for the bandwidth they share, but many of them were already paying high prices for business lines without using much of the bandwidth. However, ISF's hotspot network, which contains more than 120 locations, has given a strong community flavour to public Wi-Fi in Montreal – ISF's network is more extensive than that of any commercial provider in the city.

Beyond establishing hotspots, ISF also develops open-source software that reconfigures off-the-shelf Wi-Fi modems, transforming them into nodes in the group's network, and giving each hotspot a unique opening page with places for locally produced artwork, community content, and profiles for users logged on in any location. The idea behind the software, which is called WiFiDog,[3] is that it can create a sense of the culture of the location where someone uses Wi-Fi to connect to the Internet. Thus, by viewing images, artwork, and other information unique to a location, a user is reminded of where she/he is, of the culture of the place, and of his/her role as a citizen. WiFiDog also regulates the distribution of video and music content, and authorizes all of the users on a network. Many of WiFiDog's functionalities were developed through partnerships with arts and culture projects, indicating once again the extent to which ISF's efforts at leveraging Wi-Fi for cultural projects make it distinct.

In practical terms, WiFiDog presents an opening page that requires a new user to create an ISF account. The account must be validated using an e-mail address, and a user must log in every time he or she uses ISF services. When a user logs in, he or she sees the portal page specific to the location she/he is at; it includes news and announcements about the network as a whole, as well as specific content contributed by the manager of the hotspot. The portal page also contains a list of members on-line at that hotspot. ISF is in the process of developing user profiles that will give more information about who is on-line at each hotspot and provide people the opportunity to connect with one another. The first developers of WiFiDog hoped that their software would alter the way that people use public spaces as places for work, and would reinforce the sociability and community awareness that public spaces have historically provided. They wanted Wi-Fi to be seen as another way to be in a place with other people, rather than merely a means to access the Internet.

Making Visible the Invisible: Politicizing Technology

Do CWNs actually contribute to a different kind of understanding of Wi-Fi technology, and of the overall potential of ICTs to augment local communities? Sandvig (2004) argues that the primary missions of North American and European wireless community groups do not necessarily offer significant challenges to dominant telecommunications policy or delivery mechanisms, depending as they do on an ethics of 'accidental sharing' of bandwidth that is either already purchased from telecommunications companies or donated by an academic or industrial partner. However, as wireless technology becomes more ubiquitous, and as private companies and municipalities develop high-level, and potentially expensive, wireless Internet services, wireless groups can potentially contribute something other than frameworks for sharing signals: they can contribute by creating an alternative telecommunications infrastructure, and thus by making visible the dominance that telecommunications companies exert through their ownership of the otherwise invisible Internet infrastructure. Many CWNs were founded or organized around the idea that they could provide an infrastructure alternative to that of the increasingly commercialized Internet. Therefore, they proposed independent meshed networks that could share community information and act as a means of contacting friends and neighbours in local areas. These projects are similar to the community networking projects (see Schuler 1996) that proposed the use of computers and networks as a means to reinforce local communities.

Both meshed networks and hotspot networks can act as attempts – in different ways – to make visible and draw into question the consolidation and commercialization of Internet infrastructure. In this sense, they attempt to politicize the design of infrastructure. As Graham and Marvin (2001) point out, infrastructure is increasingly politicized as telecommunications companies increasingly avoid investment in poorer urban neighbourhoods while offering more and more diverse services in richer neighbourhoods. This brings into question the extent to which the ability to access communications infrastructure determines who is able to communicate. Extending infrastructure extends the capacity to communicate, while restricting it may well have the opposite effect. Therefore, through attempts at creating an alternate infrastructure, this power differential is brought into focus. More radically, as Mackenzie (2003) points out, the desire to analyse infrastructure brings into question the very foundations of equality:

> In many places, infrastructure no longer looks like a space of reciprocity, a space without negativity. Hence, attachment to infrastructure, the indispensable condition for a politics of infrastructure becomes contested. The link we might want to add to infrastructural imaginings concerns the very possibility of the infrastructural-political. Contestation of the collective mode of existence of infrastructure, this tenuous and still fragile fibre, is a promising development. (15)

Therefore, in a world where infrastructure (especially networked communications infrastructure) is problematically contested, politics, culture, and the physical shape of communications infrastructure become aligned.

Following Mackenzie, it is unclear whether redesigning infrastructure is sufficient as a political act. In focusing on infrastructure, we risk ignoring its fragility as a site of engagement. The development of an independent meshed network can be a specific attempt to create an alternative to the Internet, but it is technically complicated and its success depends upon the willingness of its members to develop content, and to participate in the creation and maintenance of the network. Furthermore, the purpose, stated or not, of most meshed networks seems to be to connect to the Internet. Thus their technical configuration is not in and of itself sufficient to create an alternative to the Internet. As discussed below, meshed protocols are now primarily used by large commercial enterprises as methods of covering large areas. A hotspot network can also make infrastructure visible, not by creating an alternative infrastructure but by changing the way that end users encounter the network. ISF's portal page project attempts to introduce a sensitivity for local neighbourhood culture into the use of the network, hoping to make Internet users aware of where they are and who helps them connect. However, in interviews with and observations of ISF users, most people revealed that while they occasionally consulted the portal page, they would rarely use profiles to communicate with other people using Wi-Fi in the same space. Instead, some used the list of logged-in members at each hotspot as a gauge of the bandwidth being used – and avoided locations with many logged-in members as they would have slower connection speeds.

Furthermore, despite the fact that the portal pages designed for WiFiDog were intended to create a local community at the location where the access was provided, many users picked up the signals, which 'bleed' easily through stone walls and windows, from adjacent locations. This effectively turns around one of the main goals for the ISF portal page – to reinforce local place-based communities by providing a platform for in-person exchange between people using wireless Internet at a specific location, and 'geo-specific content' linked to that location. Thus, the activities of users suggest that the ISF model of CWN, while it does to a certain extent politicize Internet infrastructure, has not succeeded in reaffirming local communities through its networking and software projects. Users seem ambivalent at best about the group's profile project, and seem most interested in getting free Wi-Fi, not in participating in a mediated version of café society.

Community Wi-Fi as a Shared Collective Imaginary

As a result of these challenges, it is tempting to conclude that ISF's creation of a CWN based on hotspots has failed to create a community supportive of its project to *make visible* local community and culture through the use of Wi-Fi. This would be a hasty conclusion. Beyond mere technical development, ISF creates the *idea* of a community alternative to commercial technology. Escobar writes, 'Any technology represents a cultural invention, in the sense that it *brings forth a world*; it emerges out of particular cultural conditions and in turn helps to create new ones' (1994, 211, my emphasis). In Montreal, ISF's efforts resonate with a culture of community action and grassroots projects. The city has a long tradition of grassroots organizing and mutual aid, extending back to the organizing efforts of the Catholic religious colonists. More recently, decades of leftist governments have solidified in citizens the concept of a 'shared good' and a connection between radical politics and community media (Raboy 1984). Therefore, the idea of a community group providing a technical service is culturally resonant. By attempting to construct a social, rather than commercial, relationship between hotspot managers, users, and the group, ISF has discursively, if not economically, restructured their relationship. From the local cultural conditions, they have also brought forth a new world – the sense of collective 'technological imaginary' and of telecommunications as a shared good. Flichy (2001) articulates his concept of the technological imaginary as being one of a shared vision for the future of a technology but also one anchored in collective action. In this sense, ISF's project has succeeded, but not in the manner that the group expected.

ISF's users, for example, were not interested in talking about how they used Wi-Fi, but they were interested in talking about ISF as a group, and their relationship to this group. They thought the project was 'a great initiative that should be reproduced elsewhere,' or that they would like to receive more information about the group. The users I interviewed described their interactions with ISF as follows: one individual described how he proposed an alternative funding model to the group; another described how using ISF made him feel 'at home,' and related how he hoped there would be more locations close to his home but did not feel capable of contacting ISF to request new locations. Another user, who had gone on to open an Internet café, maintained ISF service at his new business despite the fact that the service cost him more in lost Internet business than it gained him in new clients. He said, 'I wanted to thank ISF for all of the wonderful mornings they gave me sitting in the sun and reading the news' (Powell 2006b).

Anderson (1991) points out the extent to which national communities are 'imagined.' In this sense, the community created around the ISF services is also an imagined one, formed and strengthened through the creation of a non-profit framework for offering telecommunications systems. The idea of the non-profit organization, even more than its actual activities, helps to motivate and solidify a shared collective imaginary. In cultural terms, Wi-Fi becomes visible as a community project, a 'wonderful initiative' and a potential site for volunteers. Through ISF's logos and stickers, placed on doors and windows to identify locations with free access, Wi-Fi also becomes visible as a mark of a cultural practice and of subcultural cool that distinguishes some public locations from others. But as much as this shared collective imaginary participates in the creation of a world in which communications infrastructure might be a common good, it may not be sufficient as a form of resistance to the increasing corporate consolidation of communications infrastructure. Already, Wi-Fi is becoming visible as a municipal service, facilitated and delivered by telecommunications companies as a public service for cities.

Part II: Who Is the Community? Wi-Fi as a Public Service

The creation of a shared collective imaginary of what telecommunications infrastructure might or should be is certainly part of the world that ISF has brought forth in Montreal, but the way that Wi-Fi is conceived is already in the process of changing. Countless cities have begun to contract with large telecommunications companies to provide ubiquitous Internet access. These municipal Wi-Fi projects offer the potential for universal access to the Internet, as well as the expansion of services already offered by many municipalities, including communication between objects like parking meters and electricity meters and a city's central server (Bar and Galpernin 2004). Unlike CWN projects, municipal Wi-Fi projects can guarantee the security of data transfer, provide long-range, high-speed coverage over large areas, and use municipally owned bandwidth as a community service. However, they do not necessarily provide free Wi-Fi services, and may not necessarily be organized as non-profit organizations. North America's first operational municipal Wi-Fi network, the Fred E-zone in Fredericton, New Brunswick, is a non-profit organization and provides Wi-Fi service, based on the city's own bandwidth, free of charge in public areas. However, Fred E-zone's goal is not to use Wi-Fi to augment existing public spaces, but rather to provide a service that offers potential competition to an existing telecommunications duopoly. As Dagget (2006) points out, most other municipal wireless projects do not explicitly consider questions of ownership when requesting proposals for the development of their networks. This means that in many cities citizens will have the possibility of accessing a Wi-Fi network from anywhere, but that this network will likely be owned and operated by a telecommunications company, most likely for profit. The network itself thus becomes invisible except through a client-service relationship.

The Failure of the Collective Imaginary?

Like water systems, electrical systems, and transportation systems, to which we rarely pay attention, computer networking systems, and especially Wi-Fi, are quickly becoming infrastructures. As Bowker and Star (1999) point out, when technologies become infrastructural, they become invisible. This 'infrastructural inversion' increases the potential power of a technology, even as it disappears from view. Following this point, Mosco (2004) contrasts the 'sublime' moments of excitement around technical development with the 'banal' ones in which technologies accumulate greater power while disappearing into the background. Municipal Wi-Fi, based on service provision, marks the beginning of Wi-Fi's possibly inevitable fade into the background, and perhaps a parallel increase in its power. Sandvig (2006) likened this movement to the end of telephone cooperatives that self-organized to provide telephone service in the early twentieth century. With the advent of universal telephone service contracts, these cooperatives disbanded. Telephone use is now an everyday activity, and one for the most part that places us at the mercy of monopoly telecommunications operators. Furthermore, the lack of universal service policies for mobile telecommunications services makes communicating using these devices extremely expensive and beyond the reach of many Canadians. Little or no resistance has developed against the consolidation of influence of telecommunications companies in this sense – telephony is already invisible. Resistant or oppositional developments do not concentrate on the telephonic infrastructure so much as its potential uses.

The CWN movement has reconfigured Wi-Fi's visibility: whereas the Internet begins to 'fade into the background' in large cities, the CWN movement makes Wi-Fi visible so that the terms of its invisibility can be controlled. The attempts to make Wi-Fi a public good, or to configure it in such a way as to retain a sense of the community or culture in which it is being used, testify to a desire to see communications as a common good and to resist a blind dependence on telecommunications compan- ies. They are part of an ensemble of practices that attempt to keep sys- tems open, and to interrogate their relative visibility. By creating a collective imaginary that frames Wi-Fi as a potential public or common good, CWNs create the discursive framework that opens a space for resistance. In the ISF case, the CWN project created a framework in which the ability to communicate could be augmented as part of a vol- unteer community project. In this sense, CWNs succeed in making communications infrastructures visible, and thus in making them pol- itical. Materially, though, the moment for contesting infrastructure may have already passed. Now that municipal wireless projects are making wireless connectivity potentially ubiquitous, the place for community or collective resistance may be shrinking. Is a collective imaginary suf- ficient in and of itself as a form of resistance, considering that Wi-Fi is already becoming an invisible and highly controlled infrastructure?

Conclusion: Community in the Era of Municipal Wi-Fi Projects

There are, of course, two answers to this question: a materialistic and an idealistic. The materialistic answer, grounded in the historical cases of the telephone and the radio, suggests that the shared collective imagin- ary, no matter how powerful, endures only in the discursive realm. The real political and economic relationships underscored by the control of the material infrastructure, as well as the increasing influence of tele-com companies over these relationships, play a role too important to ignore in the definition of how we understand and use Wi-Fi. Community and collective formations have emerged around numerous technologies and systems, and have disappeared once these technologies and systems obtained stable markets. Therefore, Wi-Fi activism and the reappropria-tion of visibility of communications infrastructures will be noted in the annals of history as another peculiar phenomenon related to the rapid growth and development of a new technological system.

The idealistic answer proposes that discursive resistance is ongoing, and in fact is the most significant manner in which people make sense of their world. In this answer, the CWN project of making visible the manner in which we daily connect to the Internet, which is not only tied to a local place, space, and experience, is already a success. By estab-lishing a social relationship in place of a commercial one, ISF, in par-ticular, has established an alternative to the discourse of client and service provider. Furthermore, their explicit social mandate leaves open the possibility for their public installations to continue to provide art, information, and social opportunities in public spaces. In this answer, a resistant understanding of Wi-Fi as a carrier for social connections and as localized in a specific place might coexist with a universal telecom-munications service. Of course, the influence of the CWN model of free access to the end user would be weakened by an invisible Wi-Fi infra-structure, but not the sense that Wi-Fi might be a common or commun-ity good.

Perhaps both answers are right, each in their own measure. The long-term influence of projects like CWNs and ISF remains to be seen, but it seems likely that they are the beginning of a series of increasingly cre-ative, innovative, shifting, and dynamic tactics by which citizens and communities engage with the many overlapping layers of their cities. As cities gain new networks of influence and power, expect new net-works of resistance and reconnection – new ways of making power vis-ible – to emerge as well.

NOTES

1 Community wireless networks are groups of people, more or less organ-ized, who use wireless Internet equipment to create linkages of computers, some with connection to the Internet. Sometimes these groups have links to

other community groups, but not always. The only constants in CWNs are a non-commercial orientation and an engagement with Wi-Fi technology.

2 See http://www.picopeer.net.

3 WiFiDog, in the tradition of recursive names for free software projects, refers to a previous piece of software with the same functions – NoCat. NoCat has two mythological origins: in one, it refers to the fact that the server authorizes wireless Internet access; i.e., 'no CAT 5 cable'; in the other, it refers to Einstein's (perhaps apocryphal) explanation of wireless telegraphy: 'You see, wire telegraph is a kind of a very, very long cat. You pull his tail in New York and his head is meowing in Los Angeles. Do you understand this? And radio operates exactly the same way: you send signals here, they receive them there. The only difference is that there is no cat' (http://nocat.net).

14 'The network we all dream of': Manifest Dreams of Connectivity and Communication or, Social Imaginaries of the Wireless Commons

ANDREW HERMAN

A specter is haunting spectrum policy – the specter of the commons.
 – K. Werbach, 'Supercommons: Towards a Unified Theory of Wireless Communications,' *Texas Law Review* 82 (2004): 863

The dream of communication stops short of all the hard stuff. Sending clear messages might not make for better relations; we might like each other less the more we understood about each other. The transmission of signals is an inadequate metaphor for the interpretation of signs. 'Communication' presents itself as an easy solution to intractable human troubles: language, finitude, plurality. Why others do not use words as I do or do not feel or see the world as I do is a problem not just in adjusting the transmission and receptions of messages, but also in orchestrating the collective being, in making space in the world for each other.
 – J.D. Peters, *Speaking into the Air*, 30

What We All Dream of

In late 2001, a group of prominent American Internet policy activists penned the Wireless Commons Manifesto, which is reprinted in its entirety in this book.[1] The document was written at a time when, as Alison Powell describes in her chapter in this volume, Community Wireless Networks (CWNs) began to proliferate in urban locales across North America and Europe. The manifesto embodied a clarion call for activists of the community wireless movement to see beyond the parameters and exigencies of their local environments and to marshal their collective resources in order to establish a 'a pan-global network of local wireless communities to develop a common knowledge base and lobbying effort for spectrum reform' (http://www.nycwireless.net/tikiindex.php? page=WirelessCommons). 'We have formed the Wireless Commons,' the authors proclaimed,

> because a global wireless network is within our grasp. We will work to de-fine and achieve a wireless commons built using open spectrum, and able to connect people everywhere. We believe there is value to an independent and global network which is open to the public. We will break down com-mercial, technical, social and political barriers to the commons. The wireless commons bridges one of the few remaining gaps in universal communica-tion without interference from middlemen and meddlers ... Becoming a part of the commons means being more than a consumer. By signing your name below, you become an active participant in a network that is far more than the sum of its users. You will strive to solve the social, political and technical challenges we face. You will provide the resources your commun-ity consumes by co-operating with total strangers to build the network that we all dream of. (Wireless Commons Manifesto 2003, 386–7)

Alas, this dream remains a manifest destiny unfulfilled. The middle-men and meddlers, comprised by corporate ISPs and telecoms, remain obdurately and obstinately in place as gatekeepers and portals to the Internet backbone, especially in the United States (Sandvig 2004, Strover and Mun 2006, Meinrath 2005b, Snider 2006); the regulatory regimes that govern spectrum policy change at a seemingly glacial and piecemeal pace in the United States and Canada, and the organizations agitating and advocating for regulatory reform, are fragmented and largely in-effectual (Werbach 2004, Snider 2006, Priest in this volume); community wireless networks remain largely uncoordinated and disconnected from one another, and within these organizations there is a clear divide be-tween a small number of activists and a vast majority of subscribers who remain, first and foremost, consumers (Powell and Shade 2006); and, last but not least, the goal of a global wireless network based upon the prin-ciples of commons-based ownership and control is very far from displa-cing the hegemony of the proprietary regimes of private ownership of the means of digital connectivity and communication.[2]

Although the organizational infrastructure necessary to realize the 'dream' of the global wireless network failed to materialize in the wake of the manifesto's publication, it is nonetheless far more than a document of historical interest. To be sure, the document has not achieved the iconic status of John Perry Barlow's famous 'Declaration of Independence of Cyberspace' in the canon of cyberculture utopianism. However, the manifesto has been circulated widely on the Internet. More significantly, it is considered to be a foundational statement that defines and articulates the core values and goals of a wireless 'free culture,' a culture bounded only by availability of interoperable hardware, access to unused or unlicensed spectrum, and the strength of Wi-Fi (or Wi-Max) signal. As such, the document can be understood as a powerful performative rhetoric instantiating a particular social imaginary of a digital commons based upon wireless communication as a cultural technology of connectivity and community.

The manifesto's 'dream' of a global wireless network enabling unfettered and seemingly unmediated 'universal communication' was and is a compelling one at many levels. Yet this dream is also highly flawed in its understanding of the intertwined dynamics of communication and community, and this flaw is rooted in a particular conception of the 'commons' as a paradigm of social relations of property ownership. The rhetoric of the manifesto, I will argue, not only defines and delimits possible courses of action with respect to the regulatory policies undergirding wireless communications, but also obscures and misconstrues the social and political character of the politics of the digital communality that the wireless commons is supposed to enable. The uncritical invocation of the rhetorical trope of the 'commons,' I maintain, leads to an elision between technologies of connectivity, communicative practices, and the agonistic politics of community with potentially deleterious consequences for the enhancement of digital democracy. Thus a critical analysis of the rhetoric and discourse of the 'wireless commons' is warranted.

Lost Horizons: The Imaginary of the 'Commons,' Wireless and Otherwise

In using the terms 'imaginary' to analyse the Wireless Commons Manifesto, I am primarily drawing upon the work of Canadian philosopher Charles Taylor (1989, 2002, 2004, and Strauss 2006). For Taylor, a 'social imaginary' is a dynamic epistemological and ontological framework of cultural value and identity. Imaginaries are also more supple, and more firmly embedded in quotidian perceptions and practices, than ideologies per se. Social imaginaries, Taylor argues, entail

the ways in which people imagine their social existence, how they fit together with others, how things go on between them and their fellows, the expectations that are normally met, and the deeper normative notions and images that underlie these expectations ... I speak of *imaginary* because I'm talking about the way ordinary people 'imagine' their social surroundings, and this is often not expressed in theoretical terms; it is carried in images, stories, and legends ... the social imaginary is that common understanding that makes possible common practices and a widely shared sense of legitimacy ... This understanding is both factual and 'normative'; that is, we have a sense of how things usually go, but this is interwoven with an idea of how they ought to go, of what missteps would invalidate the practice. (2002, 106)

Social imaginaries, then, are not simply a set of *ideas* about the social world: they are pragmatic templates for social *practice*. In Foucauldian terms, imaginaries operate as forms of power-knowledge, enabling some social actions and constraining others as they provide a map of the social as moral space that is delineated along existential, normative, and utopian dimensions. That is, social imaginaries map the world in terms of what exists, what is right, and what is possible. Moreover, as Taylor argues, imaginaries are insinuated into private and public spheres of everyday life through stories and narratives. Such narratives provide a dynamic, unfolding trajectory of social being and becoming that plots a path from where we are existentially to where we want to be in terms of utopian possibility. Woven into the fabric of the story are normative evaluations of our conduct that assess how well they will get us from the former state to the latter.

At a deeper epistemological level, Ernesto Laclau maintains that an imaginary comprises a 'horizon: and, as such, the imaginary is not one among many objects but an absolute limit which structures the field of intelligibility and is thus the condition for the emergence of any object' (1990, 64). My argument here is that a peculiar conception of the 'commons' operates as the horizon of the narrative of the manifesto. This horizon renders intelligible particular existential, normative, and utopian understandings of the politics and possibilities of digital communications, wireless and otherwise. In order to understand the ways in which this imaginary limits the fields of possibility in terms of politics and policy, it is necessary to examine its structure as a narrative through which the imaginary of the commons is performed.

All narratives are given shape and direction by their teleology. In the case of the Wireless Commons Manifesto, the telos of the narrative is as a place of common identity, or *topos*, where people 'everywhere' are connected in the embrace of 'universal communication.' This *utopos* is technologically constituted by a pan-global network of interconnected local community networks that effectively annihilates geographical space and physical distance. Through this global connectivity, the ecological limits of locally based wireless signals are transcended, and strangers become potentially comrades in the utopian enterprise. As such, this *topos* embodies a realm of disembodied and disembedded virtual sociality and 'friction-free' co-presence that has long been part of the utopian mythos of cyberspace, shared equally by the likes of Bill Gates and John Perry Barlow (Herman and Sloop 2000, Mosco 2004). It matters not, as Mosco (2004, 29) has argued, whether such myths are true or false; what matters is whether they are dead or alive, and this particular myth is very much alive, and at work in the imaginary of the wireless commons.

The subject of the narrative – that is, the actor who seeks to achieve the goal – is plural or several. Within the story of the manifesto, the primary actors are the signatories, the collective 'we' of the manifesto's authorship, and the collective 'you' of future signatories and readers who find themselves interpellated by the seductive logic of the narrative's teleology. Indeed, the potential subjects of the narrative are ultimately co-extensive with all those who dream of and desire the unbounded sociality of universal communication (a desire that the authors presume is universally shared). As we shall see, as with most mythic imaginaries of the digital, the utopian place of global connectivity and universal communication operates as something of a faith-based empty signifier, evading explicit consideration of the dramaturgic correlate of communications which is community as commonality.

In narratological terms deployed by actor-network theory (Latour 1986, 1993), standing between the subjects of a story and their desired goal are two classes of actants: helpers and opponents. The actants provide the dramatic tension, as it were, giving the narrative the force to move towards its ineluctable conclusion. The opponents of the narrative's subjects are clear: standing as 'barriers to the creation of global network' are a variety of actants, technological and institutional: there are technical problems regarding the first and last mile of wireless transmission and reception that need to be resolved; there are the hardware and software vendors that need to understand the market potential of providing such technical solutions; there are the politicians and regulatory agencies who need to be persuaded to expand the spectrum available to wireless communication as well as to enable localities to pursue their own provisioning of wireless capacity quite apart from the dominant players in the marketplace, the existing telecoms and Internet Service Providers. These actors are, of course, the real villains of the story. To the authors, they are nothing more than 'limping anachronisms' that are tethered to a fate of impending obsolescence by virtue of 'the expense and physical necessity of using wires to build their networks.' Therefore, the manifesto concludes, these actors 'cannot serve the great mass of people who stand to benefit from a wireless commons. Their interests diverge from ours, and their control over the network strangles our ability to communicate.'

The primary helpers are the twin technological 'revolutions' of, first, the Internet, and, second, the emergence of broadband Wi-Fi as a relatively inexpensive means of networking that dispenses with the necessities of wired connectivity. It is concerning this crucial part of the narrative where the insights of Latour and actor-network theory become most useful in understanding the imaginary of the commons. From the perspective of actor-network theory, technologies are not passive elements in the creation of social networks; they are agents in and of themselves – not simply because of their technical properties – but because these technical properties come to be rendered meaningful to social actors through their discourses and narratives about them.

Within the narrative of the Wireless Commons Manifesto, the technologies of the Internet and Wi-Fi become powerful vectors of agency through which the telos of universal connectivity and communication will be achieved. Both embody 'revolutions' in communication and, in revolutionizing communication, have also transformed the scope and scale of human interaction and social relations. 'Humanity,' the manifesto inveighs,

> is on the verge of a turning point because the Internet has transformed the way humans relate with one another. All communication can be traced to a human relationship, whether it's lovers exchanging instant messages or teenagers sharing music. The Internet has given us the ability to communicate faster and more cheaply than ever before in history ... Low-cost wireless networking equipment which can operate in unlicensed bands of the spectrum has started another revolution. Suddenly, ordinary people have the means to create a network independent of any physical constraint except distance.

It is at this point in the narrative that the ontological individualism of the manifesto's imaginary becomes apparent. The face-to-face interaction of individuals in a singular and isolated relation of communicative inter-action and exchange becomes the urtext for 'all communication.'[3] Thus, the Internet and wireless become simply a more efficient and effective means of extending the reach of interpersonal communication and con-necting the mutual desires of *individuals in exchange relationships*, be they 'lovers' or 'teenagers,' at high speed with low expense.

A crucial part of the teleological emplotment of the story is the in-vocation of 'Reed's Law' concerning the 'value' created by network connectivity, a law apparently 'discovered' by one of the charter signa-tories, David P. Reed.[4] As expressed in the manifesto, the 'Internet's value increases exponentially with the number of people who are able to participate.' Within the context of the unfolding narrative, this means that the more people who are bound together in this digital commun-ion of individual desire expressed and individual desire fulfilled, the greater value the Internet produces. Hence, the emergence of low-cost Wi-Fi devices and networking equipment not only provides the means to evade the wired gatekeepers to the Internet, but also enhances the 'value' of the resources brought by individuals to the network.

These twin technological revolutions of Internet and wireless broad-band comprise the architectural infrastructure of what Yochai Benkler has termed the 'Networked Information Economy,' which, in turn, makes feasible a commons-based regime of the ownership and control of the means of communication and connectivity. In his magisterial (if profoundly flawed) treatise on the emergence of networked peer pro-duction and the paradigm of a commons-based approach to informa-tional property, *The Wealth of Networks*, Benkler (2006) argues that this economy is premised upon two fundamental shifts in the social econ-omy of contemporary capitalism: first, there has been the oft-theorized, analysed, and celebrated shift in economic activity in the global North from an industrial economy centred on the production of material goods to a post-industrial economy based increasingly on the produc-tion of 'immaterial' information and symbols (or what Benkler vaguely refers to as 'human meaning and communication'). Second, and more importantly in terms of the argument here, is the development of a 'pervasive network of digital communication that radically decentral-izes the capacity to make meaning and communicate' (2006, 32) and puts 'the material means of information and cultural production in the hands of a significant fraction of the world's population'(3).

In quasi-Marxist terms, Benkler argues that the development of these new means of immaterial and informational production are fettered by antiquated social relations of ownership and production embodied by proprietary regimes of intellectual property. The structural contradiction, so to speak, between the forces of production and social relations of production have produced political and juridical conflict throughout what Benkler usefully terms the 'institutional ecology of the digital environment' (2006, 23). The wireless commons is clearly an important part of this new ecology and a key terrain where the battle between proprietary and commons-based models of ownership and control over the different layers of the Internet will be enjoined.

For the authors of the manifesto, in order for the wireless network to produce the *telos* of universal communication, it *must* be governed as commons. Constructing the network as a commons is, in fact, the 'most important' element of the manifesto's imaginary:

> Most importantly, the network needs to be accessible to all and provisioned by everyone who can provide. By adding enough providers to the network, we can bridge the physical gaps imposed by the range of our equipment. The network is a finite resource which is owned and used by the public, and as such it needs to be nurtured by the public. This, by its very nature, is a commons.

It is at this point of the narrative that the horizon of the imaginary of the wireless commons as an absolute limit of possibility comes to view.

Commons with Commonality, Community without Unity

There are two dimensions of property relations that are imbricated in the understanding of the commons that is operative in the manifesto as well as in the paradigm of 'commons theory' elaborated by Benkler and others.[5] The first is the relationship of people to things or objects; the second is the relation of people to one another on the basis of their relationship of ownership and control of those things. This dual set of relations that are embodied in property is, of course, not particular to commons theory but is deeply embedded in the history of Western conceptions of property dating back to antiquity (C. Rose 1993). As David Berry and Giles Moss (2006) illustrate in their insightful genealogy of Western topologies of ownership, there are three basic forms of property relations that are pertinent here and that can be traced to Roman law and jurisprudence: *res privatae, res publicae,* and *res communes.*

Res privatae entails an exclusive property right of individuals and institutions that the law considers to be 'persons' such as corporations. This right of exclusivity entails an asymmetric power relationship between the person that owns the things and the persons that do not, namely, the former controls access by the latter to the object in question. *Res publicae* can refer either to the state ownership of things and/or the very form of resources that are owned by the state for public use (e.g., streets). In this case, the power of access and exclusivity is held by the state. Finally, in *res communes* entails a group of objects or things that are, by their very nature, either incapable of ownership or accessible to all, such as air and water. Moreover, such things that fall under the rubric of *res communes* are resources that are finite and can therefore be used up or exhausted. This particular quality requires a form of governance that controls access and manages use in order to avoid their depletion and exhaustion.

From the perspective of the manifesto, the wireless commons seems to fall clearly under the category of *res communes*; it is a finite resource that, as the public uses it as a fundamental resource of communication, should be owned and nurtured by the public. This is all very well and good, but what exactly does the rhetorical figure of the 'public' represent in this imaginary? The answer is a crucial one for, as Berry argues, the proponents of various forms of commons – such as the 'creative commons' or the 'digital commons' – often conflate common ownership with state or public ownership.[6] Historically, as Eva Hemmungs Wirtén (2006) argues, the former refers to governance by the public through customary ethics and social reciprocity, while the latter invests the power of access and exclusivity to the state *in the name of the public* (which is precisely the modus operandi of all regulatory regimes of spectrum allocation). As Berry points out, common ownership is not the same as public ownership.

Moreover, as far as the manifesto is concerned, whether the 'public' is that of *res publicae* or *res communes* is ambiguous. On the one hand, the narrative demands a veritable *potlatch* in the provisioning of the commons – all who have equipment and bandwidth contribute to the commons as if in a gift relationship of mutual obligation, and thereby nurture and govern it as resource that is held by the community in common. On the other hand, except in the abstract state of nature of what Roman law considers to be *res nullius* (the realm of unclaimed things), commons do not exist outside of the law or the state but, in fact, are constituted by them, either explicitly or in an implicit relationship of alterity. As Benkler (2006) argues:

'Commons' refers to a particular institutional form of structuring the rights to access, use, and control resources. It is the opposite of 'property' in the following sense: With property, law determines one particular person who has the authority to decide how the resource will be used. That person may sell it, or give it away, more or less as he or she pleases. However, the core characteristic of property as the institutional foundation of markets is that the allocation of power to decide how a resource will be used is systematically and drastically asymmetric. That asymmetry permits the existence of 'an owner' who can decide what to do, and with whom. We know that transactions must be made – rent, purchase, and so forth – if we want the resource to be put to some other use. The salient characteristic of commons, as opposed to property, is that no single person has exclusive control over the use and disposition of any particular resource in the commons. Instead, resources governed by commons may be used or disposed of by anyone among some (more or less well-defined) number of persons, under rules that may range from 'anything goes' to quite crisply articulated formal rules that are effectively enforced. (60–1)

Thus, contrary to what the manifesto claims, there is no such thing as a singular 'very nature' of the commons. It is always-already a political form of social organization that rests upon an explicit ethic that defines and delimits the rights and responsibilities of the commons constituents. To phrase the issue another way, the commons requires an ethic of commonality. Thus, at the very heart of the social imaginary of the wireless commons is a very large void where the social should be.

This void is nowhere more apparent in the narrative of the manifesto than in the lacuna between the provisioning of the commons as a commons and the *telos* of universal communication and connectivity. How exactly will the former enable and produce the latter? Even if we supplement the manifesto's narrative with the palimpsest of a more fully articulated understanding of the commons such as Benkler's, fundamental problems of ontological reductionism and elision remain in the imaginary of the commons, wireless and otherwise. In spite of Benkler's sociologically accurate understanding of the social relations between people and things embedded in property as being institutionally constructed, the ambiguities and ambivalences of the commonality of the commons remain. The horizon of this imaginary construes the commons fundamentally as a set of objective resources to be managed, and the social relations that emerge from the disposition of such things is assumed to naturally follow from this new disposition. As the commons is primarily a set of objective resources, the issues that concern the literature on the wireless commons are primarily technical and regulatory: At what frequency should wireless be allowed to operate? Can governmental regulatory bodies be convinced to open used 'white space' of the television broadcast spectrum to wireless? What devices allow for the maximum sharing of bandwidth in mesh networks? Can it be demonstrated that the allocation of wireless as a commons, as opposed to either licensing or auctioning off the spectrum, be a greater spur to technical innovation and creativity in wireless technologies and the provisioning of first and last mile of broadband connectivity?[7]

My point is not that these are not important questions or issues. Quite the contrary, the creation of a more democratic 'institutional ecology of the digital environment,' to use Benkler's words, requires that they must be addressed. Given that all property relations are always about social relations between people through objects, the technical constitution of those objects and their juridical disposition is crucial in the constitution of property as power. However, the commons paradigm tends to assume in reductionist fashion that the correct technical and juridical disposition of people in relation to things will, *ipso facto*, yield a communicative relationship of communality where people anywhere can connect to each other everywhere based upon an ethical sharing of communicative resources. Within the narrative, a serious conceptual elision, which bedevils the commons paradigm as a whole, becomes evident. The 'commons' as sphere of communication, the 'commons' as a place of community, and the 'commons' as a property relationship regarding technical resources become elided. A commons without either commonality or the agonistic social relations that define commonality and difference in a community produces consumers, not citizens. What is lost in this horizon is the agonistic character of the commons as a space of contingent politics where the moral character of communication and community are neither foreclosed nor guaranteed. In the end, the spectre that haunts the future of wireless is not the arrangement of spectrum as property, but the social imaginary of the wireless commons itself.

NOTES

1 The charter signatories to the manifesto include Lawrence Lessig, co-founder of the Creative Commons and one of the most well known and prolific

organic intellectuals of the digital commons movement; Cory Doctorow, noted science fiction author and, at the time of the manifesto, European representative for the Electronic Frontier Foundation; Adam Shand, one of the founders of Personal Telco, the Portland, Oregon–based community wireless network; James Stevens, of the London-based Consume CWN; Matthew Asham of BC Wireless; Paul Holman of the Shmoo Group, an organization of hackers devoted to testing and pushing the bounds of Internet security architecture; and David Reed, researcher at MIT Media Lab as well HP Laboratories, open spectrum expert and partisan, and 'discoverer' of 'Reed's Law' concerning the scalability of large networks (discussed below). The full list of the signatories can be found at http://metabolik .hacklabs.org/alephandria/txt/wirelesscommons.html.

2 In fact, in the United States there have been numerous attempts to limit municipalities' ability to provision wireless broadband as a public good; see Strover and Mun 2006.

3 This ontological reductionism, as John Durham Peters (1999) argues, has deep roots in communicative theory, going back to the pre-Socratics, and occupies a powerful yet ambivalent place in rhetorics of what Mosco (2004) has termed 'the digital sublime.'

4 Reed's Law is a revision of the so-called 'Metcalfe Law,' first formulated by the inventor of Ethernet and founder of 3Com, Robert Metcalfe. Metcalfe argued 'that the value of a network would increase quadratically – proportionately to the square of the number of its participants – while costs would, at most, grow linearly' (Briscoe, Odlyzko, and Tilly 2006, 26). More mythos than science, Metcalfe's Law was the implicit foundation of many of the suspect business models that produced the dot.com bubble of the 1990s. Some have argued that revisions and reformulations, such as Reed's, provide a potentially dangerous economic rationality for the development of many 'web 2.0' and 3G wireless innovations and applications.

5 The key canonical texts of the 'commons theory' paradigm are Lessig 2002, 2004, Bollier 2003, Boyle 2003, Rose 1993, and Hunter 2003. See Halbert 2005 for a useful sympathetic critical overview.

6 For a more developed critique of the 'Creative Commons' model, as well as the Free and Open Source Software movement, see Berry and Moss 2006.

7 Werbach (2004) provides a helpful overview of the range of technical and regulatory issues that have been central to debates around wireless spectrum policy. More current discussions of wireless technology and policy, especially in terms of the United States, can be found at the website for the 'Wireless Future' project of the New American Foundation: http://www .spectrumpolicy.org/.

Bibliography

Abbas, A. 1997. *Hong Kong, culture and the politics of disappearance*. Minneapolis: University of Minnesota Press.

Agar, J. 2005. *Constant touch: A global history of the mobile phone*. Cambridge, UK: Icon Books.

Albert, S. 2004. *Crossing the line between mapping and map making: Artistic approaches to cartography*. UO Faculty of Cartography, http://twenteenthcentury.com/saul/

Allan, S. 2002. Reweaving the internet: Online news of September 11. In *Journalism after September 11*, ed. B. Zelizer and S. Allan, 119–40. London and New York: Routledge.

Althusser, L. 1971. Ideology and ideological state apparatuses. In *Lenin and philosophy and other essays*, 121–73. London: New Left Books.

Ambulant Science, http://www.ambulantscience.org/

American Railroads. 1947. Round trip to the Moon. *LIFE*, 28 April, 9.

Amsterdam Realtime, http://www.waag.org/realtime/

Anderson, B. 1991. *Imagined communities: Reflections on the origin and spread of nationalism*. London and New York: Verso.

Anderson, C. 2006. *The long tail: Why the future of business is selling less of more*. New York: Hyperion.

Andrews, M. 2005. TV on the cell. *Leader Post*, 27 August, final edition, A1.

Angenot, M. 2004. Social discourse analysis: Outlines of a research project. *Yale Journal of Criticism* 17(2): 199–215.

Apel, D. 2004. *Imagery of lynching: Black men, white women, and the mob*. New Brunswick, NJ; London: Rutgers University Press.

Appadurai, A. 1996. *Modernity at large: Cultural dimensions of globalization*. Minneapolis and London: University of Minnesota Press.

– 2002. The right to participate in the work of the imagination. In *Transurbanism*, by Arjun Mulder, 33–48. Rotterdam: V2_Publishing.

Arceneaux, N. 2005. The world is a phone booth: The American response to mobile phones, 1981–2000. *Convergence* 11(2): 23–31.

Arendt, H. 1958. *The human condition.* Chicago: University of Chicago Press.

Augé, M. 1995. *Non-Spaces: An anthropology of supermodernity.* London: Verso, 1995.

Auray, N., C. Charbit, and V. Fernandez. 2003. WiFi: An emerging information society infrastructure. In *Socio-economic trends assessment for the digital revolution.* Milan: STAR project, IST Programme, European Commission.

Auzina, I. 2003. *Locative List,* 4 July, http://base.x-i.net/mailman/listinfo/locative (subscription required).

Ayrton, W. 1901. Synotonic wireless telegraphy. *Electrical Review,* 820.

Bakhtin, M. 1984 [1965]. *Rabelais and his world.* Bloomington: Indiana University Press.

Balint, K. 2005. Qualcomm places bets on mobile TV. *San Diego Union-Tribune,* 5 Aug., C1.

Bar, F., and H. Galpernin. 2004. Building the wireless internet infrastructure: From cordless ethernet archipelagos to wireless grids. *Communications and Strategies* 54 (2nd quarter): 45–67.

Barlow, J.P. 1996. A cyberspace independence declaration. *Electronic Frontier Foundation Publication,* http://homes.eff.org/~barlow/Declaration-Final.html

Barnett, C. 2005. Convening publics: The parasitical spaces of public action. In *The handbook of political geography,* ed. K.Cox, M. Low, and J. Robinson. London and New York: Sage. http://www.open.ac.uk/socialsciences/staff/cbarnett/Convening%20publics.pdf

Barrett, M. 1991. Ideology, politics, hegemony: From Gramsci to Laclau and Mouffe. In *The politics of truth: From Marx to Foucault,* 51–80. Stanford: Stanford University Press.

Baudrillard, J. 1983. The ecstasy of communication. In *The anti-aesthetic: Essays on postmodern culture,* ed. H. Foster, 126–34. Port Townsend, WA: Bay Press.

Bauman, Z. 2000. *Liquid modernity.* Cambridge: Polity.

Beer, G. 1996. 'Wireless': Popular physics, radio and modernism. In *Cultural babbage: Technology, time and invention,* ed. F. Spufford and J. Uglow, 149. London: Faber and Faber.

Bell Telephone Systems. 1948. Journey of a word. *LIFE.*

– 1949. The worlds in reach again. *LIFE.*

Benkler, Y. 2006. *The wealth of networks: How social production transforms markets and freedom.* New Haven, CT: Yale University Press.

Berry, D. 2006. Beyond public and private: Reconceptualising collective ownership. *Eastbound* 1(1): 153–73.

Berry, D., and G. Moss. 2006. The politics of the libre commons. *First Monday* 11(9), http://firstmonday.org/issues/issue11_9/berry/index.html

Bhabha, H. 1990. Interview with Homi Bhabha. The third space. In *Identity: Community, culture, difference*, ed. J. Rutherford, 207–21. London: Lawrence and Wishart.

Biomapping, http://www.biomapping.net/

Bollier, D. 2003. *Silent theft: The private plunder of our common wealth.* New York: Routledge.

Bolz, N.W., and W. van Reijen. 1996. *Walter Benjamin.* Trans. L. Mazzarings. Atlantic Highlands, NJ: Humanities Press.

Borgmann, A. 1984. *Technology and the character of contemporary life: A philosophical inquiry.* Chicago: University of Chicago Press.

– 1999. *Holding on to reality: The nature of information at the turn of the millennium.* Chicago: University of Chicago Press.

Bose, J.C. 1927. *Collected physical papers.* New York: Longmans, Green and Co.

Bourassa, R. 2005. Circulations alexandrines. *Intermédialités* 5: 21–36.

Bourriaud, N. 2000. *Relational aesthetics.* Dijon: Les Presses du Réel.

Bowker, G., and S.L. Star. 1999. *Sorting things out: Classification and its consequences.* Cambridge and London: MIT Press.

Boxes for baby: New style crib eliminates germs, drafts and constricting clothes. 1947. *LIFE*, 3 Nov., 73.

Boyer, M.C. 1996. *Cybercities.* Princeton: Princeton Architectural Press.

Boyle, J. 2003. The second enclosure movement and the construction of the public domain. *Law and Contemporary Problems* 66: 33–74.

Bradley, D. 2005. The divergent anarcho-utopian discourses of the Open Source Software Movement. *Canadian Journal of Communication* 3(4), http://www.cjc-online.ca/viewissue.php?id=112#Articles4

Briggs, H. 2001. Marconi's Atlantic leap remembered. BBC News Online, 11 December, http://news.bbc.co.uk/hi/english/sci/tech/default.stm

Briscoe, B., A. Odlyzko, and B. Tilly. 2006. Metcalfe's Law is wrong. *IEEE Spectrum*, July: 26–31.

Brouer, J., M. Arjen, and S. Charlton, eds. 2003. *Information is alive.* Rotterdam: V2/Nai Publishers.

Brown, B., R. Harper, and N. Green, eds. 2001. *Wireless world: Social and interactional aspects of the mobile age.* London: Springer.

Burgess, A. 2004. *Cellular phones, public fears, and a culture of precaution.* Cambridge: Cambridge University Press.

The cabled photograph is perfected: The picturegram. 1938. *LIFE*, 14 Sept., 10.

Callon, M., and J. Law. 2004. Introduction: Absence-presence, circulation, and encountering in complex space. *Environment and Planning D* 22(1): 3–11.

Canadian Radio-television and Telecommunications Commission (CRTC). 2006. *Public notice 2006–48*, 12 April.

Canadian Wireless Telecommunications Association (CWTA), http://cwta.ca/

Canetti, E. 1998 [1960]. *Crowds and power.* New York: Farrar, Straus & Giroux.

Carey, J.W. 1989. Technology and ideology: The case of the telegraph. In *Communication as culture*, 201–30. Winchester, MA: Unwin-Hyman.

Caron, A., and L. Caronia. 2007. *Moving cultures: Mobile communication in everyday life.* Montreal and Kingston: McGill-Queen's University Press.

Cartographic Command Centre, http://www.deaf04.nl/deaf04/program/events/item.sxml?uri=urn:v2:deaf04:rss:projects.rss:040929104400-ccc

Castells, M. 1996, 2000. *The rise of the network society.* Oxford: Blackwell.

Castells, M., M. Fernandez-Ardevol, J.L. Qiu, and A. Sey. 2007. *Mobile communication and society: A global perspective.* Cambridge: MIT Press.

Castoriadis, C. 1998. *The imaginary institution of society.* Trans. K. Blamey. Cambridge: MIT Press.

Chalmers, M., I. MacColl, and M. Bell. 2003. Seamful design: Showing the seams in wearable computing. *Proceedings of the IEE Eurowearable 2003*, Birmingham, UK, 11–17.

Cho, H. 2006. Explorations in community and civic bandwidth: A case study in community wireless networking. MA thesis, Communication and Culture, Ryerson University and York University.

Chun, W., and T. Keenan, eds. 2006. *New media, old media.* New York: Routledge.

Churchill, E.F., and N. Wakeford. 2002. Framing mobile collaborations and mobile technologies. In *Wireless world: Social and interactional aspects of the mobile age*, ed. R. Harper, B. Brown, and N. Green, 154–79. London and New York: Springer.

Clarke, B., and L.D. Henderson, eds. 2002. *From energy to information: Representation in science and technology, art, and literature.* Stanford: Stanford University Press.

Clement, A., and A.B. Potter. 2007. A desiderata for wireless broadband networks in the public interest. Paper presented at 35th Research Conference on Communication, Information, and Internet Policy in Arlington, VA.

Coming to a cell phone near you: Movies, sitcoms, news, sports. 2004. *Montreal Gazette*, final edition, 25 Sept., D11.

Connor, S. 2004. Topologies: Michel Serres and the shapes of thought. *Angelistik* 15: 105–17.

Cooper, A. 2006. *The hanging of Angélique: The untold story of Canadian slavery and the burning of Old Montreal.* Toronto: Haper Collins.

Cooper-Chen, A. 1997. *Mass communication in Japan.* Ames: Iowa State University Press.

Corr, O.C. 2000. *Money from thin air: The story of Craig McCaw, the visionary who invented the cell phone industry, and his next billion-dollar idea.* New York: Crown Business.

Crane, D. 2006. Cubicle dwellers' funniest home video. *New York Times*, 26 March, AR9.

Cresswell, T. 2006. *On the move: Mobility in the modern western world.* London: Routledge.

Cribb, R. 2005. TV phone yet to hit Jetsons standards. *Toronto Star*, 27 Oct., M04.

CTIA. 2006. *CTIA semi-annual wireless industry survey.* Washington, DC: International Association for the Wireless Telecommunications Industry. http://www.ctia.org

Cukier, W., and C. Middleton. 2006. Is mobile email functional or dysfunctional? Two perspectives on mobile email usage. *European Journal of Information Systems* 15(3): 252–60.

Cuneo, A.A. 2005. Marketers get serious about the 'third screen.' *Advertising Age*, 11 July, 6.

Curwen, P. 2002 *The future of mobile communications: Awaiting the third generation.* New York: Palgrave Macmillan.

CWTA. 2007. *Mobile wireless subscribers in Canada, 2005–2006.* Ottawa: Canadian Wireless Telecommunications Association. http://www.cwta.ca

Daggett, B.V. 2006. *Ownership matters.* Minneapolis: Institute for Local Self Reliance. www.newrules.org/info/wireless-ownership.pdf

Damisch, H. 2002. *A theory of /cloud/.* Trans. J. Lloyd. Stanford: Stanford University Press.

Danner, M. 2004. *Torture and truth: America, Abu Ghraib, and the war on terror.* New York: New York Review of Books.

Darroch, M. 2010. Language and the city. Forthcoming in *Circulation and the city: Essays on mobility and urban culture,* ed. A. Boutros and W. Straw. Montreal and Kingston: McGill-Queen's University Press.

Debord, G. 1958. *Theory of the dérive.* http://library.nothingness.org/articles/SI/en/display/314

Debray, R. 1996 [1994]. *Media manifestos: On the technological transmission of cultural forms.* Trans. Eric Rauth. London: Verso.

de Certeau, M. 1984. *The practice of everyday life*. Berkeley and Los Angeles: University of California Press.

de Lauretis, T. 1987. Technologies of gender. In *Technologies of gender: Essays on theory, film, and fiction*, 1–30. Bloomington: Indiana University Press.

Deleuze, G. 1992. Postscript on the societies of control. *October* 59 (Winter): 3–7.

Deleuze, G., and F. Guattari. 1983 [1972]. *Anti-Oedipus: Capitalism and schizo-phrenia*. Minneapolis: University of Minnesota Press.

Dewey, J. 1991 [1927]. *The public and its problems*. Athens: Ohio University Press.

Digital market took shape in 2005. 2006. *Toronto Star*, 21 Jan., H2.

Dixon, G. 2005. That's my film on line two. *Globe and Mail*, 10 Sept., R1.

Doss, E., ed. 2001. *Looking at LIFE Magazine*. Washington, DC: Smithsonian Press.

Douglas, S. 1987. *Inventing American broadcasting, 1899–1922*. Baltimore: Johns Hopkins University Press.

Dunne, A. 1999. *Hertzian tales: Electronic products, aesthetic experience and critical design*. London: Royal College of Art CRD Research Publications.

Electric Light and Power Company. 1945. Why did Aunt Hallie wrap the ice in paper? *LIFE*, 15 June, 25.

– 1947. How many of these need electricity? *LIFE*, 6 Oct.

Elmer, G. 2004. *Profiling machines: Mapping the personal information economy*. Boston: MIT Press.

Escobar, A. 1994. Notes on an anthropology of cyberculture. *Current Anthropology* 35(3): 211–31.

Esposito, E. 2004. The arts of contingency. *Critical Inquiry* 31(1): 7–25.

Evans, P. 2006. Movies in miniature. *Toronto Star*, 7 April, D3.

Every-ready. 1949. Every-ready mini-max: The shell with the radio brain: Army/navy lift censorship on mystery weapon that licked V-Bomb, Kamikaze attacks. *LIFE*.

The family: In western civilization it is seriously threatened and needs material and moral help. 1947. *LIFE*, 24 March, 36.

Feenberg, A. 1999. *Questioning technology*. New York: Routledge.

Fernando, A. 2006. That third screen in your pocket. *Communication World* 23(3): 11–12.

Festival terms and conditions. 2006. M*obifest*, 4 May, http://www.mobifest.ca

First annual Palm Mobifest Awards top honours to new stars of the small screen. 2006. *Canada News Wire*, 18 May, http://www.newswire.ca/en/releases/archive/May2006/17/c0145.html

Flichy, P. 2001. La Place de l'imaginaire dans l'action technique. *Réseaux* 109: 51–71.

Flusser, V. 2005. The city as wave-trough in the image-flood. Trans. P. Gochenour. *Critical Inquiry* 31(2): 320–8.

Foroohar, R. 2005. Changing channels. *Newsweek*, 6 June, 42.

Foucault, M. 1980. Truth and power. In *Power/knowledge: Selected interviews and other writings, 1972–1977*, ed. C. Gordon, 109–33. New York: Pantheon Books.

Foucault, M., and M. Blanchot. 1987. *Foucault/Blanchot*. Cambridge: MIT Press.

Frascara, J. 2003. The third skin: Wearing the car, ignoring safety. In *Mediating the human body: Technology, communication, and fashion*, ed. L. Fortunati, J.E. Katz, and R. Riccini, 195–200. Mahwah, NJ: L. Erlbaum.

Fritzsche, P. 1996. *Reading Berlin 1900*. Cambridge, MA: Harvard University Press.

Fusco, C. 2004. Questioning the frame: Thoughts about maps and spatial logic in the global present. In *These Times*, 16 Dec., http://www.inthesetimes .com/site/main/article/1750/

A fuzzy picture. 2006. *The Economist*, 7 Jan., 57.

Gabrys, J. 2007. Automatic sensation: Environmental sensors in the digital city. In *Senses and the city*. Special edition of *Senses and Society*, ed. Mags Adams and Simon Guy, 2(2): 189–200 .

– 2010. Telepathically urban. Forthcoming in *Circulation and the City*, ed. Will Straw and Alexandra Boutros. Montreal: McGill-Queen's University Press.

Gadget-loving Japanese watch TV shows on cell phones. 2006. *San Jose Mercury News*, 6 April.

Galambos, L., and E.J. Abrahamson. 2002. *Anytime, anywhere: Entrepreneurship and the creation of a wireless world*. New York: Cambridge University Press.

Galloway, A. 2004a. Playful mobilities: Ubiquitous computing in the city. Paper presented at the Alternative Mobility Futures Conference, 9–11 January Lancaster University, Lancaster, UK.

– 2004b. Intimations of everyday life: Ubiquitous computing and the city. *Cultural Studies* 18(2–3): 384–408.

Galloway, A., and M. Ward. 2006. Locative media as socialising and spatialising practice: Learning from archaeology. *Leonardo Electronic Almanac* 14(3), http://leoalmanac.org/journal/Vol_14/lea_v14_n03-04/gallowayward.asp

Gaonkar, D.P., and E.A. Povinelli. 2003. Technologies of public forms: Circulation, transfiguration, recognition. *Public Culture* 15(3): 385–97.

Gardiner, M. 2004. Wild publics and grotesque symposiums: Habermas and Bakhtin on dialogue, everyday life and the public sphere. In *After Habermas: New perspectives on the public sphere*, ed. J. Roberts and N. Crossley, 28–48. Oxford: Blackwell.

General Electric. 1948. Electronics park. *LIFE*, 19 Jan.

Gertzen, J. 2005. The latest tech irony: Wireless cable. *Kansas City Star*, 3 Nov., A1.

Giddens, A. 1998, 2000. *The third way: The renewal of social democracy.* Cambridge, UK, and Malden, MA: Polity Press.

Giesecke, M. 2002. Literature as product and medium of ecological communication. *Configurations* 10(1): 11–35.

Gillette Blue Blades. 1947. Speaking of speed, Chalmers 'Slick' Goodlin uses Gillette Blue Blades. *LIFE*, 2 Jan., 32.

Gitelman, L. 2006. *Always already new: Media, history, and the data of culture.* Cambridge: MIT Press.

Glotz, P., S. Bertsch, and C. Locke, eds. 2005. *Thumb culture: The meaning of mobile phones for society.* Bielefeld: Transcript Verlag.

Gnoffo, T. 2005. TV anytime, anywhere. *Philadelphia Inquirer*, 14 Oct., A01.

Goggin, G. 2006. *Cell phone culture: Mobile technology in everyday life.* New York: Routledge.

Gomes, P. 2004. *Locative List*, 14 May, http://base.xi.net/mailman/listinfo/locative (subscription required).

Goodall, P. 1995. *High culture, popular culture: The long debate.* St Leonards: Allen and Unwin.

Gow, G., and R. Smith. 2006. *Mobile and wireless communications.* Maidenhead, UK, and New York: Open University Press.

Graham, S. 2004a. *The cybercities reader.* London and New York: Routledge.

– 2004b. *Software-sorted geographies.* Durham University e-Prints, http://eprints.dur.ac.uk/archive/00000057/

Graham, S., and S. Marvin. 2001. *Splintering urbanism: Networked infrastructures, technological mobilities, and the urban condition.* London and New York: Routledge.

Grant, G. 1995. *Philosophy in the mass age.* Toronto: University of Toronto Press.

Green, N. 2003. Outwardly mobile: Young people and mobile technologies. In *Machines that become us: The social context of personal communication technology,* ed. J.E. Katz, 201–17. New Brunswick, NJ, and London: Transaction Publishers.

Green, S., P. Harvey, and H. Knox. 2005. Scales of place and networks: An ethnography of the imperative to connect through information and communications technologies. *Current Anthropology* 46(5): 805–26.

Greenfield, A. 2006. *Everyware: The dawning age of ubiquitous computing.* New York: New Riders.

Griffin, M. 1996. Literary studies +/– literature: Friedrich A. Kittler's media histories. *New Literary History* 27(4): 709–16.

Grover, R. 2005. Your favorite TV show is calling. *Business Week*, 17 January, 37, http://www.businessweek.com/magazine/content105-03/b3916407_mz011.htm

Guillory, J. 2004. The memo and modernity. *Critical Inquiry* 31(1): 108–32.

Gumbrecht, H.U. 2004. *Production of presence: What meaning cannot convey.* Stanford: Stanford University Press.

Gupta, A., and J. Ferguson. 1997. *Anthropological locations: Boundaries and grounds of a field science.* Berkeley and Los Angeles: University of California Press.

Gye, L. 2002. Who's watching whom? Monitoring and accountability in mobile relations. In *Wireless world: Social and interactional aspects of the mobile age*, ed. R. Harper, B. Brown, and N. Green, 32–45. London and New York: Springer.

– 2007. Picture this: The impact of mobile camera phones on personal photographic practices. *Continuum: Journal of Media & Cultural Studies* 21(2): 279–88.

Habermas, J. 1989 [1962]. *The structural transformation of the public sphere: An inquiry into a category of bourgeois society.* Cambridge: MIT Press.

Halbert. D.J. 2005. *Resisting intellectual property.* New York: Routledge.

Hall, J. 2004. Mobile entertainment: The power of play. *The Feature*, 16 June, http://www.thefeature.com/article?articleid=100764&ref=7960944

Hall, S. 1990. The whites of their eyes. (Revised.) In *The media reader*, ed. M. Alvarado and J.O. Thompson, 7–23. London: BFI.

Hamill, L., and A. Lasen, eds. 2005. *Mobile world: Past, present and future.* London: Springer.

Hamilton, A. 2007. Wireless street fight. *Time*, 26 Feb., 38.

Hannam, K., M. Sheller, and J. Urry. 2006. Mobilities, immobilities and moorings. *Mobilities* 1(1): 1–22.

Hannay, A. 2005. *On the public.* London: Routledge.

Haraway, D. 1997. *Modest_Witness@Second_Millenium.FemaleMan©_Meets_OncoMouse™: Feminism and Technoscience.* New York: Routledge.

– 1998. Deanimations: Maps and portraits of life itself. In *Picturing science, producing art*, ed. C.A. Jones and P. Galison, 181–210. New York and London: Routledge.

Harkin, J. 2003. *Mobilisation: The growing public interest in mobile technology.* London: Demos.

Heckman, D. 2006. 'Do you know the importance of a skypager?': Telecommunications, African Americans, and popular culture. In *The cell phone reader: Essays in social transformation*, ed. A. Kavoori and N. Arceneaux, 173–86. New York: Peter Lang.

Heidegger, M. 1966. *Discourse on thinking.* Trans. J. Anderson and E.H. Freund. New York: Harper and Row.

– 1971. *Poetry, language, thought*. Trans. A. Hofstadter. New York: Harper and Row.

– 1977. *The question concerning technology and other essays*. Trans. William Lovitt. New York: Harper Torchbooks.

– 1998. Traditional languages and technological language. Trans. W. Gregory. *Journal of Philosophy Research* 23: 129–45.

Hein, K., and S. McClelland. 2005. Nielsen: Cell users are tuning out TV. *Brandweek*, 31 Oct., 4.

Heiser, J. 2005. Good circulation. *Frieze* 90: 79–83.

Hemment, D. 2006. Locative arts. *Leonardo* 39(4): 348–55.

Henkin, D.M. 1998. *City reading: Written words and public spaces in antebellum New York*. New York: Columbia University Press.

Herman, A., and J. Sloop. 2000. 'Red Alert!': Heaven's Gate and friction free capitalism. In *The World Wide Web and contemporary cultural theory: Magic, metaphor, and power*, ed. Thomas Swiss and Andrew Herman, 77–98. New York: Routledge.

Heyer, P., and D. Crowley. 1995. Introduction. In *The bias of communication*, by H. Innis, ix–xxvi. Toronto: University of Toronto Press.

Hills, M. 2002. *Fan cultures*. London and New York: Routledge.

Hjelm, J. 2000. *Designing wireless information services*. London: John Wiley & Sons.

Hoete, A., ed. 2004. *ROAM: Reader on the aesthetics of mobility*. London: Black Dog Publishing.

Hong, S. 2001. *Wireless: From Marconi's black-box to the audion*. Cambridge: MIT Press.

Horst, H. 2006. The blessings and burdens of communication: Cell phones in Jamaican transnational social field. *Global Networks: A Journal of Transnational Affairs* 6(2): 142–60.

Horst, H., and D. Miller. 2006. *The cell phone: An anthropology of communication*. Oxford: Berg.

Hugo, V. 1831. *The hunchback of Notre Dame*, book five, chapter 2, http://www.classicreader.com/read.php/sid.1/bookid.330/sec.24/

Hunter, D. 2003. Cyberspace as place, and the tragedy of the digital anticommons. *California Law Review* 91: 439–518.

IEEE pervasive computing. 2004. http://www.computer.org/portal/site/pervasive/menuitem.e7bfeea1f36bd84da84840898bcd45f3/index.jsp?&pName=pervasive_level1&path=pervasive/content&file=about.xml&xsl=article.xsl&

Île Sans Fil. 2003. Île Sans Fil, English, www.ilesansfil.org (accessed 21 April 2005).

Ito, J. 2004. Emergent democracy. In *Extreme democracy*, ed. M. Ratcliffe and J. Lebokowsky, http://www.extremedemocracy.com/

Ito, M. 2005. Introduction. In *Personal, portable, pedestrian: Mobile phones in Japanese life*, ed. M. Ito, D. Okabe, and M. Matsuda, 1–16. Cambridge: MIT Press.

– ed. 2002. *Medeia no bunka no kenryokusakuyo (Creating rights in a media culture)*. Tokyo: Serika shobo.

Ito, M,. and D. Okabe. 2005. Technosocial situations: Emergent structuring of mobile e-mail use. In *Personal, portable, pedestrian: Mobile phones in Japanese life*, ed. M. Ito, D. Okabe, and M. Matsuda, 257–77. Cambridge: MIT Press.

Ito, M., D. Okabe, and M. Matsuda, eds. 2005. *Personal, portable, pedestrian: Mobile phones in Japanese life*. Cambridge: MIT Press.

Ives, N. 2004. Marketers are about to aim at the third screen: The one on the cell phone in your pocket. *New York Times*, 8 Nov., C8.

Jinnai, H. 1995. *Tokyo: A spatial anthropology*. Berkeley: University of California Press.

Jones, A. 2007. DNA 'bar-coding' could reveal untold new species, scientists say. *Globe and Mail*, 19 Feb., A7.

J-Wave Editorial Group, ed. 1993. *Medeia Sedai no karuchyaashiin (The culture scene of the media generation)*. Tokyo: J-Wave Publications Shinkoosha.

Kahn, D., and G. Whitehead, eds. 1994. *The wireless imagination: Sound radio and the avant-garde*. Cambridge: MIT Press.

Kaika, M., and E. Swyngedouw. 2000. Fetishizing the modern city: The phantasmagoria of urban technological networks. *International Journal of Urban and Regional Research* 24(1): 120–38.

Kan: Rekishi.Kamkyoo.Bunmei (Kan: History, environment, civilization). 2005. *Johoo to wa nanika? (What is communication?)*. Volume 20. Tokyo: Fujihara shoten.

Kang, J., and D. Cuff. 2005. Pervasive computing: Embedding the public sphere. *Washington and Lee Law Review* 62: 93–147.

Kasza, G.J. 1993. *The state and the mass media in Japan 1918–1945*. Berkeley: University of California Press.

Kato, F., et al. 2005. Uses and possibilities of the *Keitai* camera. In *Personal, portable, pedestrian: Mobile phones in Japanese life*, ed. M. Ito, D. Okabe, and M. Matsuda, 301–10. Cambridge: MIT Press.

Katz, J. 1997. Social and organizational consequences of wireless communications: A selective analysis of residential and business sectors in the United States. *Telematics and Informatics* 14(3): 233–56.

– 2006. *Magic in the air: Mobile communication and the transformation of social life*. New Brunswick, NJ: Transaction Publishers.

Katz, J., and M. Aakhus, eds. 2002. *Perpetual contact: Mobile communication, private talk, public performance.* Cambridge: Cambridge University Press.

Katz, J., and P. Aspden. 1999. Mobile communications: Theories, data, and potential impact. In *Connections: Social and cultural studies of the telephone in American life,* ed. James E. Katz, 41–73. New Brunswick, NJ, and London: Transaction Publishers; Berkeley: University of California Press.

Kavoori, A., and N. Arceneaux. 2006. *The cell phone reader: Essays in social transformation.* New York: Peter Lang.

Kember, S. 1998. *Virtual anxiety: Photography, new technologies and subjectivity.* Manchester: Manchester University Press.

Kennedy, R. 2006. The shorter, faster, cruder, tinier, TV show. *Time Magazine,* 28 May, 44–9.

Kerschbaumer, K. 2003. RealOne goes mobile. *Broadcasting & Cable* 133(33): 22.

Kierkegaard, S. 1978 [1846]. *Two Ages: The age of revolution and the present age: A literary review.* Princeton: Princeton University Press.

Kittler, F. 1991. *Discourse networks.* Stanford: Stanford University Press.

Knauer, L.M. 2001. Eating in Cuban. In *Mambo montage: The latinization of New York,* ed. A. Lao-Montes and A. Davila, 425–47. New York: Columbia University Press.

Kojima, K., and H. Yoshiaki. 1998. *Kawaru medeia to shakai seikatsu (Changing media and social life).* Tokyo: Minerubua Shoten.

Kolko, B.E., L. Nakamura, and G.B. Rodman, eds. 2000. *Race in cyberspace.* New York: Routledge.

Kopomaa, T. 2000. *The city in your pocket: Birth of the mobile information society.* Helsinki: Gaudeamus.

– 2002. Mobile phones: Place-centered communication and new-community. *Planning Theory and Practice* 3(2): 241–5.

Krauss, R. 1999. *A voyage on the North Sea: Art in the age of the post-medium condition.* London: Thames and Hudson.

Kwon, M. 2002. *One place after another: Site-specific art and locational identity.* Cambridge and London: MIT Press.

Lacey, L. 2005. You can be a potato anywhere. *Globe and Mail,* 5 Nov., R1, R12.

Laclau, E. 1990. *New reflections on the revolution in our time.* London: Verso.

Lane, G. 2004. Social tapestries: Public authoring and civil society. *Proboscis: Cultural Snapshots* 9, http://proboscis.org.uk/publications/SNAPSHOTS_socialtapestries.pdf

Lane, G., C. Brueton, G. Roussos, N. Jeremijenko, G. Papamarkos, D. Diall, D. Airantzis, and K. Martin. 2006. Public authoring & feral robotics. *Proboscis: Cultural Snapshots* 11, http://proboscis.org.uk/publications/SNAPSHOTS_feralrobots.pdf.

Larsen, J., J. Urry, and K. Axhausen, eds. 2006. *Mobilities, networks, geographies.* Aldershot, UK: Ashgate.

Lash, S. 2002. Informational totemism interview, by A. Mulder. In *Transurbanism,* by A. Mulder, 49–63. Rotterdam: V2_Publishing.

Latour, B. 1986. *Laboratory life: The construction of scientific facts.* Princeton, NJ: Princeton University Press.

– 1987. *Science in action. How to follow scientists and engineers through society.* Cambridge: Harvard University Press.

– 1993. *We have never been modern.* Cambridge: Harvard University Press.

– 2005. From realpolitik to dingpolitk, or how to make things public. In *Making things public: Atmospheres of democracy,* ed. B. Latour and P. Weibel, 14–41. Cambridge: MIT Press.

Latour, B., and P. Weibel, eds. 2005. *Making things public.* Karlsruhe: ZKM.

Leading edge foot traffic. 2006. *New Zealand Marketing Magazine,* 5 April.

Lee, B., and E. LiPuma. 2002. Cultures of circulation: The imaginations of modernity. *Public Culture* 14(1): 191–213.

Lefebvre, H. 1991. *The production of space.* Oxford and Malden: Blackwell Publishers.

Lessig, L. 2002. *The future of ideas: The fate of the commons in a connected world.* New York: Vintage.

– 2004. *Free culture: The nature and future of creativity.* New York: Penguin.

Leung, J. 2006. Broadcast your fandom: An analysis of fan produced concert videos, music fan culture and YouTube.com. Unpublished paper, York University, Toronto.

Leung, L., and R. Wei. 1999. Who are the mobile have nots? Influences and consequences. *New media and society* 1(2): 209–26.

Levinson, P. 2004. *Cellphone: The story of the world's most mobile medium and how it has transformed everything!* New York: Palgrave McMillan.

Levy, S. 2005. Television reloaded: The transformation is underway. *Newsweek,* 6 June, 60.

Lewis, N. 2005. The cell-ver screen: Mobile phone movies ring in change. *Calgary Herald,* 5 Oct., C1, final edition.

Lichty, P. 2004. *Building a culture of ubiquity,* http://www.voyd.com/ubiq/

Ling, R. 2004. *The mobile connection: The cell phone's impact on society.* Oxford: Morgan Kauffman.

Ling, R., and L. Haddon. 2003. Mobile telephony, mobility, and the coordination of everyday life. In *Machines that become us: The social context of personal communication technology,* ed. J. Katz, 245–65. New Brunswick, NJ, and London: Transaction Publishers.

Ling, R., and P. Pedersen. 2005. *Mobile communications: Re-negotiation of the social sphere.* London: Springer.

Ling, R., and B. Yttri. 2002. Hyper-coordination via mobile phones in Norway. In *Perpetual contact: Mobile communication, private talk, public performance*, ed. J. Katz and M. Aakhus, 139–69. Cambridge: Cambridge University Press.

Lippmann, W. 1925. *The phantom public*. New York: Harcourt Brace.

Location, Location, Location, http://www.eventnetwork.org.uk/petegomes/

Locative List, http://base.x-i.net/mailman/listinfo/locative

Lovink, G. 2005. *The principle of networking: Concepts in critical Internet culture*. Amsterdam: HvA Publicaties.

Lowenthal, D. 1985. *The past is a foreign country*. Cambridge: Cambridge University Press.

Luckhurst, R. 2002. *The invention of telepathy*. Oxford: Oxford University Press.

Lycett, J.E., and R.I.M. Dunbar. 2000. Mobile phones as Lekking devices among human males. *Human nature* 11(1): 93–104.

Lynch, L., and E. Razlogova. 2006. The Guantanamobile project. *Vectors 2*, http://www.vectorsjournal.org/index.php?page=7&projectId=3

Lyon, D. 2003. *Surveillance after September 11*. Oxford: Blackwell.

– ed. 2006. *Theorizing surveillance: The panopticon and beyond*. Cullompton: Willan.

Mackenzie, A. 2003. The infrastructural-political. *M/C: A Journal of Media and Culture* 6(4), http://journal.media-culture.org.au/0308/05-infrastructural.php. (accessed 1 May 2006).

– 2005. Untangling the unwired: Wi-Fi and the cultural inversion of infrastructure. *Space and Culture* 8(3): 269–85.

Making mobile movies. 2006. *Mobifest*, 4 May, http://www.mobifest.ca

Manly, L. 2006. For tiny screens, some big dreams. *New York Times*, 21 May, BU1, BU4.

Manovich, L. 2001. *The language of new media*. Cambridge: MIT Press.

– 2005. *The poetics of augmented space: Learning from Prada*, http://www.manovich.net/DOCS/augmented_space.doc.

Marcus, G. 1998. *Ethnography through thick & thin*. Princeton: Princeton University Press.

Markis, S. 2005. Cell 'mobisodes' capture consumer imagination. *Calgary Herald*, 15 March, D10.

Markoff, J., and M. Fackler. 2006. With a cellphone as my guide. *New York Times*, 28 June, www.nytimes.com (accessed 16 Feb. 2007).

Marres, N. 2005. Issues spark a public into being: A key but often forgotten point of the Lippmann-Dewey debate. In *Making things public: Atmospheres of democracy*, ed. B. Latour and P. Weibel, 208–17. Cambridge: MIT Press.

– 2006. Public (im)potence. In *Open 11: Hybrid space: How wireless media mobilize public space*, ed. J. Seijdel, 78–81. Rotterdam: NAi Publishers.

Marvin, C. 1990 [1988]. *When old technologies were new: Thinking about electronic communication in the late nineteenth century*. Oxford: Oxford University Press.

McCarthy, A. 2001. *Ambient television*. Durham, NC: Duke University Press.

McCullough, M. 2004. *Digital ground: Architecture, pervasive computing, and environmental knowing*. Cambridge: MIT Press.

McDonough, T. 1996. The derive and situationist Paris. In *Situationists: Art, politics, urbanism*, ed. L. Andreotti and S. Costa, 54–66. Barcelona: MACBAR.

McLuhan, M. 1994a. Automation: Learning a living. In *Understanding media: The extensions of man*, 347. Cambridge: MIT Press.

– 1994b. Telegraph: The social hormone. In *Understanding media: The extensions of man*, 257. Cambridge: MIT Press.

– 1994c. *Understanding media: The extensions of man*. Cambridge: MIT Press.

McNicoll, T. 2005. A world of digital dim sum. *Newsweek*, international edition, 26 Sept., 94.

Meinrath, S. 2005a. Wirelessing the world: The battle over (community) wireless networks. In *The future of the media: Resistance and reform in the 21st century*, ed. Robert McChesney, 219–42. New York: Seven Stories Press.

– 2005b. Community wireless networks, participatory media, and neighborhood empowerment. Presentation at Institute for Communications Research Brownbag Series, University of Illinois, Urbana-Champaign, Illinois.

Merchant, C. 1983. *The death of nature: Women, ecology, and the scientific revolution*. New York: Harper and Row.

Merriden, T. 2003. *Rollercoaster: The turbulent life and times of Vodafone and Chris Gent*. Oxford: Capstone Publishing Ltd.

Milk, http://milkproject.net/

Milroy, S. 2003. The pleasures of heavy metal. *Globe and Mail*, 24 May, R6.

– 2006. From fun all the way to fearsome. *Globe and Mail*, 11 July, R1.

Milutis, J. 2006. *Ether: The nothing that connects everything*. Minneapolis: University of Minnesota Press.

Miniature wrist radio: US Bureau of Standards develops a Dick Tracy wrist transmitter which can broadcast messages for a mile. 1947. *LIFE*, 6 Oct., 61.

Mitchell. W. 2004. *Me++: The cyborg self and the networked city*. Cambridge: MIT Press.

Mons, A. 2002. *La traversée du visible: Images et lieux du contemporain*. Paris: Editions de la Passion.

Moretti, F. 2005. *Graphs maps trees: Abstract models for a literary theory*. London: Verso.

Morley, D. 2000. *Home territories: Media, mobility and identity*. London: Routledge.

Morley, D., and K. Robins. 1995. *Spaces of identity: Global media, electronic landscapes and cultural boundaries.* London and New York: Routledge.

Mosco, V. 2004. *The digital sublime: Myth, power and cyberspace.* Cambridge: MIT Press.

Motion sickness: Willing student take rough rides in experiment to find out what makes travelers get nauseated. 1947. *LIFE,* 12 Feb., 54.

Motorola. 1944. Errol Flynn and Motorola radio: Handie-talkie land with paratroopers in 'Objective Burma.' *LIFE.*

Mulder, A. 2002. *Transurbanism.* Rotterdam: V2_Publishing.

Murmur, http://murmur.ca/

Murphy, C.J.V. 1947. The last 500 feet: With new inventions the airman is piercing the final weather barrier. *LIFE,* 12 Dec., 82.

Murray, J.B. 2001. *Wireless nation: The frenzied launch of the cellular revolution in America.* Cambridge, MA: Perseus Publishing.

Musgrove, M. 2005. Proteus teams with ABC to offer cell phone content. *Washington Post,* 9 Aug., D04.

MUTE. 2005. 2(1). Underneath the knowledge commons.

Nafus, D., and K. Tracey. 2002. Mobile phone consumption and concepts of personhood. In *Perpetual contact: Mobile communication, private talk, public performance,* ed. J.E. Katz and M. Aakhus, 206–21. Cambridge and New York: Cambridge University Press.

Neto, I., M.L. Best, and S.E. Gillett. 2005. License-exempt wireless policy: Results of an African survey. *Information Technologies and International Development* 2(3): 73–90.

New York skyline photo. 1947. *LIFE,* 31 March (picture of the week).

Nicholson, J. 2008. Calling Dick Tracy! Or cellphone use, progress, and a racial paradigm. *Canadian Journal of Communication* 33(3): 379–404.

Noblis-Parks Industries. 1949. ARVON. New super–powered portable really reaches out! *LIFE,* 2 May, 30.

No CRTC regulation planned for phone-a-vision. 2006. *Edmonton Journal,* 13 April, G2.

Noguchi, Y. 2006. CBS to make a soap for the smaller screen. *Washington Post,* 12 Jan., D5.

Nowlin, S. 2005. The third screen. *San Antonio Express-News,* 8 Jan., 1C.

Ostherr, K. 2005. *Cinematic prophylaxis: Discourse and contagion in the discourse of world health.* Durham, NC: Duke University Press.

Parks, L.B. 2005. Please pass the popcorn. *Houston Chronicle,* 28 Jan., 1.

Parviainen, J. 2002. Bodily knowledge: Epistemological reflections on dance. *Dance Research Journal* 34(1): 11–26.

Peters, J.D. 1999. *Speaking into the air: A history of the idea of communication.* Chicago: University of Chicago Press.

Pharr, S., J. Krauss, and S. Ellis. 1996. *Media and politics in Japan.* Honolulu: University of Hawaii's Press.

Philco. 1944a. The first network! Another milestone in the progress of television (Philadelphia-NY Schenectady). *LIFE.*

– 1944b. From radar research to radio for your home. *LIFE.*

– 1947. Amazing new Philco auto radios. *LIFE,* 17 Nov.

Pinck, P. 2000. From the sofa to the crime scene: Skycam, local news and the televisual city. In *Urban space and representation,* ed. M. Balshaw and L. Kennedy, 55–68. London: Pluto Press.

Plant, S. 2002. *On the mobile: The effects of mobile telephones on social and individual life.* London: Motorola.

– 2003. Mobile knitting. In *Information is alive,* ed. J. Brouer, A. Mulder, and S. Charlton. Rotterdam: V2/Nai Publishers.

Pogue, D. 2006. TV here, there, everywhere. *New York Times,* 23 March, C1, C10.

Pope, S. 2005. The shape of locative media. *Mute magazine* 29.

Popular Science. 1973 [July]. Cover. Retrieved 11 June 2007 from http://www.lumenelle.com/about/ken_larson.html

Powell, A. 2006a. Île Sans Fil as a digital formation, final report: Pratiques collaboratives. LBCO, École des Médias, Université du Québec à Montréal.

– 2006b. Last mile or local innovation? Community Wi-Fi as civic participation. Paper presented at Telecommunications Policy Research Conference, 29 Sept.–2 Oct, Arlington, VA.

Powell, A., and L.R. Shade. 2006. Going WiFi in Canada: Municipal and community initiatives. *Government Information Quarterly* 23(3–4): 381–403.

Pullman. 1947. Let's go! And here's how to get there! Go Pullman. *LIFE,* 28 April, 25.

Puro, J. 2002. Finland: A mobile culture. In *Perpetual contact: Mobile communication, private talk, public performance,* ed. J.E. Katz and M. Aakhus, 19–29. Cambridge and New York: Cambridge University Press.

Raboy, M. 1984. *Movements and messages: Media and radical politics in Quebec.* Toronto: Between the Lines.

Radio hat: Two tube set plays fine but looks ridiculous. 1949. *LIFE,* 6 June, 20.

Radio pictures. 1939. *LIFE,* 13 Feb., 22.

Rafael, V.L. 2003. The cell phone and the crowd: Messianic politics in the contemporary Philippines. *Public Culture* 15(3): 399–425.

Raiford, L. 2003. The consumption of lynching images. In *Only skin deep: Changing visions of the American self,* ed. C. Fusco and B. Wallis, 267–73.

New York: International Center of Photography and Harry N. Abrams Publishers.

Rajiva, L. 2005. *The language of empire: Abu Ghraib and the American media*. New York: Monthly Review Press.

Rakow, L.F. 2002. *Gender on the line: Women, the telephone, and community life*. Urbana: University of Illinois Press.

Rakow, L.F., and V. Navarro. 1993. Remote mothering and the parallel shift: Women meet the cellular telephone. *Critical Studies in Mass Communication* 10(2): 144–57.

Rancière, J. 2006. Problems and transformations in critical art. In *Participation*, ed. C. Bishop, 83–93. London and Cambridge, MA: Whitechapel MIT Press.

Razack, S.H. 2005. How is white supremacy embodied? Sexualized racial violence at Abu Ghraib. *Canadian Journal of Women and the Law* 17(2): 341–63.

RCA. 1947a. RCA: Radio Corporation of America RCA Laboratories – your magic carpet to new wonders of radio and television. *LIFE*, 27 Oct., 12.

– 1947b. RCA: Radio Corporation of America RCA Radar – enables ships to see through fog, darkness, storms. *LIFE*, 14 July, 83.

– 1947c. Teleran – radio eyes for blind flying (teleran pictures – air traffic control by radar plus television! *LIFE*.

– 1947d. FM Radio – another world in listening pleasure! *LIFE*, 17 March, 12.

– 1947e. New FM – noiseless as the inside of a vacuum tube! *LIFE*, 20 Oct., 26.

Regulator to keep hands off mobile TV. 2006. *Toronto Star*, 13 April, A28.

Rheingold, H. 2002. *Smart mobs: The next social revolution*. Cambridge: Perseus Publishing.

Rifkin, A. 1993. *Street noises: Parisian pleasure 1900–40*. Manchester: Manchester University Press.

Rivière, C. 2005. Mobile camera phones: A new form of 'being together' in daily interpersonal communication. In *Mobile communications: Re-negotiation of the social sphere*, ed. R. Ling and P.E. Pedersen, 167–85. London: Springer-Verlag.

Robbins, B., ed. 1993. *The phantom public sphere*. Minneapolis: University of Minnesota Press.

Robbins, K.A., and M.A. Turner. 2002. United States: Popular, pragmatic and problematic. In *Perpetual contact: Mobile communication, private talk, public performance*, ed. J.E. Katz and M. Aakhus, 80–93. Cambridge and New York: Cambridge University Press.

Robertson, G. 2008. BCE tangled up in TV fee fight. *Globe and Mail*, 9 April, http://www.theglobeandmail.com/servlet/story/RTGAM.20080409 .wcrtcfollow0409/BNStory/robNews/

Robinson, W., and D. Robison. 2006. Tsunami mobilizations: Considering the role of mobile and digital communication devices, citizen journalism, and the mass media. In *The cell phone reader: Essays in social transformation*, ed. A. Kavoori and N. Arceneaux, 85–103. New York: Peter Lang Publishing.

Rogers, R. 2004a. Why map? The techno-epistemological outlook. In *Transcultural mapping*. Acoustic space #5, ed. M. Tuters and R. Smite. Riga: Center for New Media Culture RIXC. http://www.locative.net/tcmreader/index.mapping;rogers

– 2004b. *Information politics on the web*. Cambridge: MIT Press.

Romano, A., and K. Kerschbaumer. 2005. The accidental journalist. *Broadcasting & Cable* 135(28): 17.

Ronell, A. 1989. *The telephone book: Technology, schizophrenia, electric speech*. Lincoln: University of Nebraska Press.

Rony, F.T. 1996. *The third eye: Race, cinema, and ethnographic spectacle*. Durham, NC: Duke University Press.

Roos, J.P. 1993. Sociology of cellular telephone: The Nordic model (300,000 yuppies?). *Telecommunictions Policy* 17(6): 446–58.

Rose, C. 1993. *Property and persuasion: Essays in the history, theory and rhetoric of ownership*. Boulder, CO: Westview.

Rose, G. 1993. *Feminism and geography: The limits of geographical knowledge*. Minnesota: University of Minnesota Press.

Rose, N. 1999. *Powers of freedom: Reframing political thought*. Cambridge: Cambridge University Press.

Russel, B. 2002. *Headmap ... location aware devices*, http://www.headmap.org/

Sacchi, L. 2004. *Tokyo: City and architecture*. New York: Universe.

Sandvig, C. 2004. An initial assessment of cooperative action in WiFi networking. *Telecommunications Policy* 28(7–8): 579–602.

– 2006. Network neutrality is the new common carriage. Paper presented at the Public Service Telecommunications Conference, University of Illinois at Urbana-Champaign.

Sant, A. 2004. Redefining the basemap. In *Trans-cultural mapping*. Acoustic space #5, ed. M. Tuters and R. Smite. Riga: Center for New Media   Culture RIXC. http://www.locative.net/tcmreader/index.php?mapping;sant

Sassen, S. 2001. *The global city: New York, London, Tokyo*. Princeton: Princeton University Press.

Savage, M., and A. White. 2005. TU Media in world first for mobile TV test run. Media Asia, 28 January, http://www.brandrepublic.com/News/4600171/

Schivelbusch, W. 1979. *The railway journey: Trains and travel in the 19th century New York*. Trans. Anselm Hollo. New York: Urizen Books.

Schmidt, T., and A. Townsend. 2003. Why wireless networks want to be free. *Communications of the Association for Computing Machinery (ACM)* 46(5): 47–52.

Schuler, D. 1996. *New community networks: Wired for change.* New York: Addison-Wesley.

Sconce, J. 2000. *Haunted media: Electronic presence from telegraphy to television.* Durham, NC: Duke University Press.

Seijdel, J., ed. 2006. *Open 11: Hybrid space: How wireless media mobilize public space.* Rotterdam: NAi Publishers.

Serres, M. 1982. *The parasite.* Trans. Lawrence R. Scher. Baltimore: Johns Hopkins University Press.

Shade, L. 2007. Feminizing the mobile: Gender scripting of mobiles in North America. *Continuum: Journal of Media and Cultural Studies* 21(2): 179–89.

Shaw, G. 2005. Telus wants you to watch where you're going. *Vancouver Sun,* 14 May, G1.

Shecter, B. 2006. Televolution. *National Post,* 25 March, FP1, FP6.

Sheller, M. 2004. Mobile publics: Beyond the network perspective. *Environment and Planning D: Society and Space* 22: 39–52.

Sheller, M., and J. Urry. 2003. Mobile transformations of 'public' and 'private' life. *Theory, Culture & Society* 20(3): 107–25.

– eds. 2006. *Mobile technologies of the city.* London: Routledge.

Sherry, J., and T. Salvador. 2002. Running and grimacing: The struggle for balance in mobile work. In *Wireless world: Social and interactional aspects of the mobile age,* ed. R. Harper, B. Brown, and N. Green, 108–19. London and New York: Springer.

Shields, M. 2005. Content providers see phones as '3rd Screen'. *Adweek,* 4 April.

Shields, R. 1997. Flow as a new paradigm. *Space and Culture* 1: 1–8.

Shockfish, S.A. 2008. SpotMe website. On-line at www.spotme.com. (accessed 9 May 2008).

Shunya, Y. 1994. *Medeia jidai no bunka shakaigaku (Socio-cultural studies in the media age).* Tokyo: Shinyosha.

Shunya, Y., and K.S. Jung. 2001. *Guroobaruka no enkinhoo: Atarashii kookyoo kukan o motomete (Perspectives on globalization: In search of new public space).* Tokyo: Iwanami shoten.

Siegert, B. 1999. *Relays: Literature as an epoch of the postal system.* Trans. K. Repp. Stanford: Stanford University Press.

Siklos, R. 2006. Can TV's and PC's live together happily ever after? *New York Times,* 14 May, BU3.

Silver, D. 2000. Margins in the wires: Looking for race, gender and sexuality in the Blacksburg electronic village. In *Race in cyberspace,* ed. B.E. Kolko, L. Nakamura, and G.B. Rodman, 133–50. New York: Routledge.

Simmel, G. 2000 [1903]. The metropolis and mental life. In *The city cultures reader*, ed. M. Miles, T. Hall, and I. Borden, 12–19. New York: Routledge.

Situations and Imaginary Journey, http://www.mlab.uiah.fi/~htikka/

Smith, J. 2004. Good network needed to watch TV on cell phone. *Rocky Mountain News*, 13 Sept., 12B.

Snider, J. 2006. Spectrum policy wonderland: A critique of conventional property rights and commons theory in a world of low power wireless devices. Presented at the Telecommunication Policy Research Conference, George Mason University School of Law, and Arlington, VA, 30 Sept.

Sobchack, V. 1996. Introduction: History happens. In *The persistence of history: Cinema, television, and the modern event*, ed. V. Sobchack, 1–14. New York and London: Routledge.

Soja, E.W. 1996. *Thirdspace: Journeys to Los Angeles and other real-and-imagined places*. Oxford: Blackwell.

Solanas, F., and O. Getino. 1983. Towards a third cinema. In *Twenty-five years of new Latin American cinema*, ed. Michael Chanan, 17–27. London: British Film Institute.

Spigel, L. 1988. Installing the television set: Popular discourses on television and domestic space, 1948–1955. *Camera Obscura* 16: 11–47.

– 1992. *Make room for TV: Television and the family ideal in postwar America*. Chicago: University of Chicago Press.

Steinbock, D. 2003. *Wireless horizon: Strategy and competition in the worldwide mobile marketplace*. New York: Amacon.

– 2005. *The mobile revolution: The making of mobile services worldwide*. London: Kogan Page Ltd.

– 2006. Advent of mobile TV offers big branding options to pioneers. *Media Asia*, 23 February, http://www.redorbit.com/news/technology/403170/advent_of_mobile_tv_offers_big_branding_to_pioneers/index.html

Sterling, B. 2005. *Shaping things*. Cambridge: MIT Press.

Stewart-Warner. 1947. Home means so much more when the whole family enjoys this thrilling entertainment! Wonder Window Television. *LIFE*, 121.

Stierle, K. 1993, 2001. *La capitale des signes: Paris et son discours*. Paris: Editions de la Maison des sciences de l'homme. Trans. M. Rocher-Jacquin. Original publication: *Der Mythos von Paris. Zeichen und Bewußtein der Stadt*. Munich, Vienna: Carl Hanser Verlag.

Strauss, C. 2006. The imaginary. *Anthropological Theory* 6(3): 322–44.

Strover, S., and S.H. Mun. 2006. Wireless broadband, communities and the shape of things to come. *Government Information Quarterly* 23(3–4): 348–58.

Suchman, L. 2000. *Located accountabilities in technology production.* Lancaster, UK: Centre for Science Studies, Lancaster University. http://www.comp. lancs.ac.uk/sociology/papers/Suchman-Located-Accountabilities.pdf

Surtees, L. 1994. *Wire wars: The Canadian fight for competition in telecommunications.* Scarborough, ON: Prentice-Hall Canada.

Suzuki, A. 2001. *Do android crows fly over the skies of an electronic Tokyo?* London: AA Publications.

Taylor, B. 2006. The down side of no negative. *Toronto Star,* 1 Oct., D1, D11.

Taylor, C. 1989. *Sources of the self: The making of the modern identity.* Cambridge: Harvard University Press.

– 2002. Modern social imaginaries. *Public Culture* 14(1): 91–124.

– 2004. *Modern social imaginaries.* Durham, NC: Duke University Press.

Taylor, K. 2005. This year the revolution has been digitized. *Globe and Mail,* 24 Dec., R4.

– 2006. With TV hitting cell phones, can Cancon survive? *Globe and Mail,* 22 April, R7.

Taylor, W.R. 1991. Broadway: The place that words built. In *Inventing Times Square: Commerce and culture at the crossroads of the world,* ed. W.R. Taylor, 212–31. New York: Russell Sage Foundation.

Terepurezensu, Intaakomyuunikeeshion Tokushu (Telepresence, Inter-communications Special Issue), No. 25. 1998. Tokyo: NTT Shuppan.

Terranova, T. 2000. Free labor: Producing culture for the digital economy. *Social Text* 18(263): 33–57.

Tesla, N. 1915. The wonder world to be created by electricity. *Manufacturer's Record,* 9 Sept., http://www.pbs.org/tesla/res/res_art03.html

– 1919. The true wireless. *Electrical Experimenter,* May, http://www.pbs.org/ tesla/res/res_art06.html

– 1999 [1915]. *My inventions: The autobiography of Nikola Tesla.* Ed. B. Johnston. New York: Barnes & Noble Books.

Thrift, N. 2004. Intensities of feeling: Towards a spatial politics of affect. *Geografiska Annaler* 86 B (1): 55–76.

Thrift, N., and S. French. 2002. The automatic production of space. *Transactions of the Institute of British Geographers* 27(3): 309–35.

Tikka, H. 2005. Syntymiä/births mobile service experiment – agency in participatory productions. Paper presented at the Agency in Technically Mediated Society seminar, Helsinki Institute of Science and Technology Studies, 6–8 Sept.

Tiny camera: They are becoming popular gadgets for kids, sportmen and snoopers. 1949. *LIFE,* 20 Jan., 51–2.

Toffler, A. 1981. *The third wave.* Toronto: Bantam Books.

Toranzuaato T., ed. 2003. *Hon to konpyuutaa (TransArt, special issue, The book and the computer)*. Tokyo: Dainihon inshoo kabushiki kaisha.

Tourists: Man-made spectacles rival scenery as an attraction. 1947. *LIFE*, 7 April 134.

Townsend, A. 2000. Life in the real-time city: Mobile telephones and urban metabolism. *Journal of urban technology* 7(2): 85–104.

– 2004. Mobile communications in the twenty-first century. In *Wireless world: Social and interactional aspects of the mobile age*, ed. B. Brown, N. Green, and R. Harper, 62–77. London: Springer-Verlag.

Trans Western Airline (TWA). 1947. Airline radar is here! *LIFE*, 2 June.

Trav-ler. 1947. Travel with Trav-ler: Listen ... and be carried away! *LIFE*, 3 June.

Turner, P., and E. Davenport, eds. 2005. *Spaces, spatiality and technology*. Dordrecht: Springer.

Tuters, M. 2004. *The locative commons: Situating location-based media in urban public space*, http://www.futuresonic.com/futuresonic/pdf/Locative_Commons.pdf

Tuters, M., and R. Smite, eds. 2004. *Trans-cultural mapping. Acoustic space #5*. Riga: Center for New Media Culture RIXC.

Ubiquity Interactive. 2004. *Research report: VUEGuide*. Vancouver: Ubiquity Interactive.

– 2006a. *Marketing report: VUEGuide*. Vancouver: Ubiquity Interactive.

– 2006b. *Making the city clickable*. Ubiquity Interactive press release, 12 May.

University of Openess, Faculty of Cartography. 2004. *The London free map: Your right to open geographic data*, http://locative.us/freemap/propaganda.pdf

Urban Tapestries, http://urbantapestries.net/

Urry, J. 2000. *Sociology beyond societies: Mobilities for the twenty-first century*. London: Routledge.

Virilio, P. 1994. *The vision machine*. Bloomington: Indiana University Press.

Virno, P. 2004. *A grammar of the multitude*. Los Angeles and New York: Semiotext(e).

Walk, http://www.socialfiction.org/dotwalk/

Walsh, J. 2004. Exploring the Eurospatial cartel. In *Trans-cultural mapping. Acoustic space #5*, ed. M. Tuters and R. Smite. Riga: Center for New Media Culture RIXC. http://www.locative.net/tcmreader/.

Warchalking, http://www.blackbeltjones.com/warchalking/index2.html

Wark, M. 1994. *Virtual geography: Living with global media events*. Bloomington: Indiana University Press.

Warner, M. 2002. *Publics and counterpublics*. New York: Zone Books.

Warschauer, M. 2000. Language, identity and the Internet. In *Race in cyberspace*, ed. B.E. Kolko, L. Nakamura, and G.B. Rodman, 151–70. New York: Routledge.

Wellbery, D.E. 1990. Foreword. In *Discourse Networks 1800/1900*, ed. Friedrich A. Kittler, trans. Michael Metteer, with Chris Cullens, vii–xxxiii. Stanford: Stanford University Press.

Werbach, K. 2004. Supercommons: Towards a unified theory of wireless communications. *Texas Law Review* 82: 863–973.

Werts, D. 2006. Unglued from the tube: The changing ways we watch TV. *Newsday*, April, C16.

Western Electric. 1947. 'Radar puts the finger on our enemies!' (Western electric. In peace … source of supply for bell system; in war… arsenal of communications equipment. *LIFE*.

Western Union Telegram. 1947. New radio wings can speed 2000 telegrams at once. *LIFE*, 20 Jan., 23.

White, B. 2004. FCC chief opposes wireless regulation. *Fort Worth Star Telegram*, 23 March, 2C.

Whitney, A. 2006. Can you fear me now? Cell phones and the American horror film. In *The cell phone reader: Essays in social transformation*, ed. A. Kavoori and N. Arceneau, 125–38. New York: Peter Lang.

Whyte, M. 2005. Films for the cineplex that's in your purse. *Toronto Star*, 9 Aug., C5.

Williams, M. 2006. RFID fitted throughout Tokyo neighbourhood. *PC Advisor*, 26 Dec. On-line at www.pcadvisor.co.uk (accessed 7 Feb. 2007).

Williams, R. 1977. Structures of feeling. In *Marxism and Literature*, 128–35. Oxford: Oxford University Press.

– 1992. [1974]. *Television: Technology and cultural form*. Hanover: Wesleyan University Press.

Wilson, K. 2008. The last mile: Service tiers versus infrastructure development and the debate on Internet neutrality. *Canadian Journal of Communication* 33(1): 81–100.

Wireless Commons Manifesto. 2003. *Sarai Reader* 03: 386–7, http://www.wirelesscommons.org

Wireless TV efforts recognized. 2006. *Calgary Herald*, final edition, 11 April, C12.

Wirtén, E.H. 2006. Out of sight and out of mind: On the cultural hegemony of intellectual property (critique). *Cultural Studies* 20(2–3): 282–91.

Wright, G. 1983. *Building the dream: A social history of housing in America*. Cambridge: MIT Press.

Zelizer, B. 2002. Photography, journalism, and trauma. In *Journalism after September 11*, ed. B. Zelizer and S. Allan, 48–68. London; and New York: Routledge.

Contributors

Darin Barney
*Canada Research Chair in Technology and Citizenship; Associate Professor,
Art History and Communication Studies, McGill University*
Darin Barney is the author of *Communication Technology: The Canadian
Democratic Audit* (UBC Press, 2005), *The Network Society* (Polity Press,
2004), and *Prometheus Wired: The Hope for Democracy in the Age of Network
Technology* (UBC/Chicago/UNSW, 2000). He is also co-editor with
Andrew Feenberg of *Community in the Digital Age: Philosophy and
Practice* (Rowman and Littlefield, 2004).

Sandra Buckley
Research Fellow, Australian National University, School of Art
Sandra Buckley is a leading researcher in the ethics and politics of mobile communication technologies. Her writing and teaching have developed around a long research relationship with Japanese and East
Asian culture, with a strong focus on practices at the margins of mainstream media and cultural and knowledge production. She recently co-
edited a special issue of the journal *Fibreculture* on the topic of mobile
technologies. Dr Buckley has also enjoyed research fellowships at the
Canadian Centre for Architecture, Banff Centre for the Arts, Humanities
Research Centre at the Australian National University, and the Centre
for Cultural Studies UCSC, among others. Dr Buckley was Director of
the Centre for the Humanities and Performing Arts at SUNY-Albany,
Chair of Japanese Studies at Griffith University, and Chair of East Asian
Studies at McGill University. She is one of the founding editors of the
path-breaking series *Theory Out of Bounds*, published by University of
Minnesota Press. Her publications include *Broken Silence: Voices of*

Japanese Feminism (University of California Press, 1997) and *Encyclopedia of Contemporary Japanese Culture* (Routledge, 2002).

Barbara Crow
Associate Professor / Technology Enhanced Learning Chair, York University
Barbara Crow is the Associate Dean of Research in the Faculty of Liberal Arts and Professional Studies at York University. She co-directs the Mobile Media Lab. Research projects have included: 'Community Wireless Infrastructure Research Project,' examining wireless fidelity as infrastructure; 'Digital Cities,' focusing on the relationship between digital technology and multimedia cities; 'Canadian Sexual Assault Law and Contested Boundaries of Consent: Legal and Extra-Legal Dimensions,' with Lise Gotell, investigating women's organizations and legal discourses; and, most recently, 'Mobile Digital Commons Network,' exploring relations of mobile technologies and cultural production. She was president of the Canadian Women's Studies Association, 2002-4.

Jennifer Gabrys
Lecturer, Design and Sustainability, Goldsmiths College, University of London
Jennifer Gabrys is currently completing a book-length study on electronic waste, *The Natural History of Electronics*. Essays she has written include 'Media in the Dump,' published in *Trash: Alphabet City Magazine* (2006); 'Appliance Theory,' published in *Cabinet Magazine* (2006); and 'Airdrop,' published through *Bookworks* (2004). She has been a researcher and contributing writer with the Digital Cities Project and Mobile Digital Commons Network since 2003.

Anne Galloway
Senior Lecturer, School of Design, University of Wellington
Anne Galloway is a social researcher working at the intersection of technology, space, and culture. Anne's PhD was on the social and cultural dimensions of mobility, and the design of mobile and pervasive technologies for urban public spaces. Anne's research has been presented to international audiences in technology, design, art, architecture, and social and cultural studies, and her work has been published in academic journals and industry magazines, as well as discussed in the popular press. When not pursuing her own research, Anne enjoys teaching undergraduate courses in social studies of science and tech-

nology and critical cultural theory. In her spare time, Anne can be found hanging out with her cat and reading comics.

Kajin Goh
Project Designer, Digital Cities, Mobile Digital Commons Network
Kajin Goh currently lives and works in Vancouver, dividing his time between design and independent art practice. Projects include founding the graffiti art and urban culture magazine *Under Pressure*, architectural design for the Urban Gallery – a prototype 'free-zone' for graffiti writers in Montreal – and graphic material for Concordia University, the Canadian Centre of Architecture, Ellen Lupton, Naomi Klein, and old-school breakdance outfit the Rock Steady Crew (NY). He has participated in the following group shows: Out for Fame, a graffiti art/photography exhibit; and Interstate (VAV Gallery, Montreal). His video work has been acquired by Concordia University for the Contemporary Art History program.

Andrew Herman
Associate Professor, Communication Studies, Wilfrid Laurier University
Andrew Herman's current research interests reside at the interdisciplinary interstices of medium theory, Internet studies, mobile communications technologies, and the politics of intellectual property. He is the author of many scholarly publications, including *The Better Angels of Capitalism: Rhetoric, Narrative and Moral Identity of the American Upper Class* (Westview, 1998; nominated for the C. Wright Mills Award in 1999) and articles in such journals as *Cultural Studies, Critical Studies in Media Communications, Theory and Society, South Atlantic Quarterly, Media and Society,* and the *Anthropological Quarterly.* He is also editor of *Mapping the Beat: Popular Music and Contemporary Cultural Theory* (Blackwell, 1997) and *The World Wide Web and Contemporary Cultural Theory* (Routledge, 2000). He is currently completing work on a new book on the politics of intellectual property in digital contexts with Rosemary J. Coombe, Canada Research Chair at York University.

Michael Longford
Associate Professor, Design, York University
Michael Longford's creative process and research activities converge at the intersection of photography, graphic design, and new media. Working with a multi-disciplinary research group, he recently completed an exploratory project, *Imaging Digital Cities,* examining the deployment of

a communications technologies infrastructure in the urban environment. His latest international symposium, *Declarations of Interdependence and the Immediacy of Design*, explored the public sphere as a space of democratic voice and citizenship in the context of socially engaged design practice. He is a founding member of Hexagram, was the Research Director for the Advanced Digital Imaging and 3D Rapid Prototyping Group, and co-directs the Mobile Media Lab.

Judith A. Nicholson
Assistant Professor, Communication Studies, Wilfrid Laurier University
Judith A. Nicholson has published articles and reviews on mobile communication in *Canadian Journal of Communication*, *Fibreculture Journal*, *M/C: A Journal of Media and Culture*, and *Topia: Canadian Journal of Cultural Studies*.

Alison Powell
Post-doctoral Fellow, Oxford Internet Institute
Alison Powell holds a post-doctoral fellowship from the Social Sciences and Humanities Research Council as a Visting Fellow at the Oxford Internet Institute. She has a PhD in Communication Studies from Concordia University in Montreal. Her work focuses on the politics and culture of community wireless groups, and on the design, governance, and regulation of urban Wi-Fi networks. She is also interested in the practices of technological development, especially the development of wireless technologies for use in public spaces. She is a member of the Canadian Research Alliance on Community Innovation and Networking (CRACIN), where she examines community and municipal wireless projects. She is also a member of the LabCMO research group on computer-mediated communication at Université du Québec à Montréal. She has presented and published on the relationships between ICTs, citizenship, and public space, as well as on the emergence of community technology groups in Canada. Her latest work, 'Going WiFi in Canada: Municipal and Community Initiatives,' co-authored with Leslie Regan Shade, appeared in 2006 in *Government Information Quarterly*. In the summer of 2006 she completed a project on 'public,' 'private,' and 'shared' Wi-Fi at the École Nationale Supérieure des Télécommunications in Paris.

Julian Priest
Independent researcher and artist
Julian Priest was co-founder of the early wireless free-network community Consume.net in London. He became an activist and advocate for the free-networking movement and has pursued wireless networking as a theme in the fields of art, development, and policy. He has written and lectured extensively on the subject through Informal – an informal independent research framework. He was one of the instigators of WSFII, the world summits on free information infrastructures, which was an international series of events to promote grassroots information infrastructures.

He has commented on radio spectrum policy and co-founded the policy intervention OpenSpectrum UK to advocate an open spectrum in the public interest, in Europe and the UK. Since 2005 he has developed an artistic practice around participatory and collaborative forms and has shown works internationally in the UK, Latvia, Germany, and New Zealand. He has worked with students and been peer advisor at the Banff New Media Institute in Canada.

He is currently focused on art practice, and his current interests are around the physical and cultural boundaries between technology and the environment. Most recently these have found expression in a new show, 'A Geekosystem.' He is based in Whanganui, New Zealand, where he has opened a new public project room, 'The Green Bench.'
Sites: http://consume.net, http://picopeer.net, http://informal.org.uk, http://wsfii.org, http://openspectrum.org.uk, http://geekosystem.org, and http://greenbench.org.

Kim Sawchuk
Professor, Communication Studies, Concordia University
Kim Sawchuk is the current editor of the *Canadian Journal of Communication*. She has recently completed a special issue for the journal on digital communication technologies and practices entitled 'Life on Line.' Dr Sawchuk has written numerous articles on feminism, technology, medicine, and the new media. She is the co-editor of *When Pain Strikes* (1999, with Cathy Busby and Bill Burns) and *Wild Science: Reading Feminism, Medicine, and the Media* (2000, co-edited with Janine Marchessault) and is currently involved in a co-production with

Dr Christina Lammer from the Medical University of Vienna entitled *Patient Knowledge*. She has been the Director of the Joint PhD in Communications at Concordia University and the Director of the Masters in Media Studies Program. She is on several editorial boards, including *Topia* and *Feminist Media Studies*. In 1996 she co-founded StudioXX, a research and digital media arts centre for women in Montreal. She co-directs the Mobile Media Lab.

Will Straw
Department of Art History and Communications Studies, McGill University
Will Straw is the author of over fifty articles on film, music, and urban culture. He is the co-editor of several volumes, including *The Cambridge Companion to Pop and Rock* and *Accounting for Culture: Thinking through Cultural Citizenship*. Dr Straw is on the editorial boards of several journals, including *Screen, Convergences, Cultural Studies, Culture, Theory and Critique*, and *Space and Culture*. He is the co-editor, with Alexandra Boutros, of *Circulation and the City: Essays on Mobility and Urban Culture* (Montreal and Kingston: McGill-Queen's University Press, 2009).

Minna Tarkka
Director, m-cult: centre for new media culture, Helsinki
Minna Tarkka has authored and edited numerous articles and publications on media, design, and contemporary culture, as well as produced media art and design projects for museums, television, on-line, and wireless environments. She was Professor of Interactive and Multimedia Communication at the Media Lab, University of Art and Design, Helsinki (1996–2001) and a senior researcher at the National Consumer Research Centre (2002–3). She was the first director of MUU artist's association (1989–90) and program director of the ISEA94 Helsinki symposium (1993–4). She has acted as a new media expert on Finnish and Nordic councils and committees, and is a founding member of m-cult: centre for new media culture. Her research aims at a critical study of new media, participation, and governmentality.

Digital Futures is a series of critical examinations of technological development and the transformation of contemporary society by technology. The concerns of the series are framed by the broader traditions of literature, humanities, politics, and the arts. Focussing on the ethical, political, and cultural implications of emergent technologies, the series looks at the future of technology through the 'digital eye' of the writer, new media artist, political theorist, social thinker, cultural historian, and humanities scholar. The series invites contributions to understanding the political and cultural context of contemporary technology and encourages ongoing creative conversations on the destiny of the wired world in all of its utopian promise and real perils.

Series Editors:
Arthur Kroker and Marilouise Kroker

Editorial Advisory Board:
Taiaiake Alfred, University of Victoria
Michael Dartnell, University of New Brunswick
Ronald Deibert, University of Toronto
Christopher Dewdney, York University
Sara Diamond, Banff Centre of the Arts
Sue Golding (Johnny de philo), University of Greenwich
Pierre Levy, University of Ottawa
Warren Magnusson, University of Victoria
Lev Manovich, University of California, San Diego
Marcos Novak, University of California, Los Angeles
John O'Neill, York University
Stephen Pfohl, Boston College
Avital Ronell, New York University
Brian Singer, York University
Sandy Stone, University of Texas, Austin
Andrew Wernick, Trent University

Books in the Series:
Arthur Kroker, *The Will to Technology and the Culture of Nihilism: Heidegger, Nietzsche, and Marx*
Neil Gerlach, *The Genetic Imaginary: DNA in the Canadian Criminal Justice System*
Michael Strangelove, *The Empire of Mind: Digital Piracy and the Anti-Capitalist Movement*
Tim Blackmore, *War X: Human Extensions in Battlespace*
Michael Dartnell, *Insurgency Online: Web Activism and Global Conflict*

Janine Marchessault and Susan Lord, eds., *Fluid Screens, Expanded Cinema*
Barbara Crow, Michael Longford, and Kim Sawchuk, eds., *The Wireless Spectrum: The Politics, Practices, and Poetics of Mobile Media*